What people are saying about

Transhuman Citizen

There are only a few people in the world who are experts on how futurist issues and libertarian philosophy collide. Dr Ben Murnane is one of them. His personal journey—battling a rare genetic illness, Fanconi anemia, and low life-expectancy—also gives him a unique and urgent perspective on life-extension and transhumanist issues. After introducing transhumanism to hundreds of millions of people through my writings and political campaigns, I was looking for a biographer to objectively capture how my activism has upended the status quo of the futurist world. Dr Murnane was the perfect fit to objectively write this story for the first time.

Zoltan Istvan

Transhuman Citizen

Zoltan Istvan's Hunt for Immortality

Ben Murnane

Previous Books

Two in a Million: A True Story about Illness and Love
ISBN 978-1-906353-03-2
Dublin on a Shoestring (with Katherine Farmar)
ISBN 978-1-906353-10-0
Feather Silence: Poems ISBN 978-1-906353-24-7
Ayn Rand and the Posthuman: The Mind-Made Future
ISBN 978-3-319-90852-6

Transhuman Citizen

Zoltan Istvan's Hunt for Immortality

Ben Murnane

CHANGEMAKERS
BOOKS

Winchester, UK
Washington, USA

CollectiveInk

First published by Changemakers Books, 2024
Changemakers Books is an imprint of Collective Ink Ltd.,
Unit 11, Shepperton House, 89 Shepperton Road, London, N1 3DF
office@collectiveinkbooks.com
www.collectiveinkbooks.com
www.changemakers-books.com

For distributor details and how to order please visit the 'Ordering' section on our website.

Text copyright: Ben Murnane 2023

ISBN: 978 1 80341 528 4
978 1 80341 554 3 (ebook)
Library of Congress Control Number: 2023936016

A CIP catalogue record for this book is available from the British Library.

Design: Lapiz Digital Services

UK: Printed and bound by CPI Group (UK) Ltd, Croydon, CR0 4YY
Printed in North America by CPI GPS partners

Contents

Anyone who fights for the future, lives in it today.
Ayn Rand

Acknowledgements

First and foremost, I'd like to thank Zoltan, for approaching me to work on this project, and for committing so much time to share his story and meticulously fact-check my work. It's been a thoroughly enjoyable journey.

Second and foremost, thank you to my wife, Sandra Cronin, who listened to me talk about this book for years, as I worked in whatever spare time I had to complete it. Sandra offered so much invaluable feedback along the way.

A huge thank-you to every interviewee and contact who spared time to make this volume better, including Chris T. Armstrong, Lincoln Cannon, Pratik Chougule, Rachel Edler, Ilona Gyurko, Gabriella Gyurko Ashford, Gary Johnson, Lisa Memmel, and Gennady Stolyarov II.

To my parents, Mai and Des, and sisters, Ruth and Jess, for their continuing support.

To John Douglas, Emma Eager, and Mai, who read over the finished manuscript and gave advice that made it better.

If I've forgotten anyone above... you know, still, you are not forgotten.

Prologue

"Should a Transhumanist Run for US President?"

With that rhetorical question, Zoltan Istvan announced he was campaigning to be America's Commander-in-Chief. His latest *Huffington Post* blog—with that question as its headline— was going viral. A lot of people, clearly, were excited about a candidate who would finally put science and technology at the forefront of the political conversation...

Zoltan's number-one campaign promise was, he wrote, "to do everything possible to make it so this country's amazing scientists and technologists have resources to overcome human death and aging within 15–20 years—a goal an increasing number of leading scientists think is reachable."

He had thrown down the gauntlet to the reader and to every American—to every person in the world, in fact:

> In the next 25 years, all of us will face a choice about how far we want to take technology and science—all of us will face a Transhumanist Wager. Artificial hearts will become better than the best human hearts. Bionic arms will become superior to human arms. Smart phones will become the size of a fingernail and will likely be implanted into your body. Speaking out loud will disappear as the modern world uses mindreading headsets to communicate, which already exist. Where will you stand? How far will you take technology in your life?[1]

Will you oppose these radical changes, or embrace them?

Immigration, terrorism, the performance of the economy, even the threat of climate change—none of these was the biggest political issue of the day, according to Zoltan. In fact, all these "problems" could be solved with advanced technology, if we decided to pursue it.

Surveillance of immigrants and terrorists with drones... Advanced robots and AI that keep the engines of economic production running, allowing humans to live off the robots' labor... Tech that gives us unparalleled mastery over nature: to control the climate, to live in space, even to recreate our own bodies. All these things are possible with transhumanism; we just have to work toward them.

No mainstream politician is willing to talk about the real existential issue of our century, according to Zoltan.

Are we willing to bet everything—to invest the maximum resources, to marshal support all around the globe—to ensure the Transhumanist Age comes about?

For Zoltan, it was about one very simple drive: With the power of these new things, we could halt aging. We could stop death. We could live forever. And not just live forever—live in new forms and experience what was currently impossible... That was the kind of science-fiction existence he wanted for himself.

He'd already become a bestselling author on these subjects... an in-demand speaker and commentator on podcasts and on TV and in newspapers all around the world. But he needed to do something even bigger, on a bigger stage, create even more of a buzz.

Now, this California native felt ready to go out there and take the fight to every corner of America, on the presidential campaign trail.

He knew there was bound to be opposition to his "crazy" ideas at every turn. From fellow transhumanists... from religious conservatives... From all quarters, people would oppose how he framed the world and the fight for the decades to come. He had to expect it.

After all, Zoltan intended to provoke conversations, debates, about the largest topics of all time: the nature of humanity and the future of everything.

But Zoltan Istvan Gyurko had always fought to chart his own course. Whether it was defying his parents as a young man to sail around the globe... or flying into war zones as a journalist when others were fleeing... or spotting a business opportunity that made him a fortune in just a few years... he didn't care much about others' expectations. Only creating his own future.

He was ready to take his message to the biggest stage in the world.

Chapter 1

A Pioneer's Life

Zoltan grimaced.

His 26-foot sailing boat, *The Way*, was alone on the open sea. The water had grown dark and choppy.

A storm was coming—a big one. He hadn't faced a challenge like this before. It was early in 21-year-old Zoltan's attempted circumnavigation of the globe. He'd left Hawaii and was on his way to the island nation of Kiribati. The Pacific Ocean was all that could be seen in every direction. There were hundreds of miles before he reached a port.

The winds were growing intense now. He did the rounds and double-checked the rigging. Loose rigging sinks ships. Once the boat was as prepared as it could be, there was little he could do except stay alert. He could wait out most of the storm in the cabin, but he'd have to surface and check *The Way* often—spot trouble early: damage to the hull, a crack in the mast... Jesus, it could be bad. He popped two caffeine pills. He'd need energy in the hours to come.

People spoke idly about the virtues of nature. About wanting to live in an unspoiled environment; how humans were destroyers and untouched nature was perfection. They rarely considered that nature was the greatest killer in existence. Whether it was earthquakes, viruses, beasts in the wild, or old age, only human ingenuity had blunted the effects of deadly nature. With early-warning systems, vaccines, guns, and pills. In other words, with science and technology. The human mind was our only defense against nature, the killer.

The same law applied to sailing. There was no escaping nature on a sailboat. You were up close against the elements in every moment. Your preparedness and quick thinking could save you. A misjudgment or just bad luck could doom you. For Zoltan to enjoy every glorious sunset from the deck of his "floating apartment," he also had to face the hurricanes.

To watch a storm coming in... there was a terrifying beauty in it. To see the world shifting color, blue to white to gray to black—each shade a slip closer to destruction. And to be alone in the expanse and face nature this way: him and his mortality. It was a thrill. It was what he'd signed up for in this life the moment he defied his parents and escaped to Costa Rica, at age 16, spending two months smoking weed and getting high on mushrooms, surfing and hitch-hiking along the coast. For as long as he could remember, he'd wanted to be an adventurer, a pioneer, to push the boundaries. It was the same psychology, in a way, that had led him to sell drugs (he'd spent a month in jail for selling weed, just before he took to the ocean)... To take every drug he could (once using LSD 27 days in a single month)... To climb tall buildings without a rope (a favorite pastime in college, until he was kicked out after his drug arrest)... The same psychology that would soon lead him to surf down erupting volcanoes (launching a sport called "volcano boarding" on the National Geographic Channel), and take him into war zones...

What was life without living it on the edge?

Maybe he got this attitude from his parents. As he tells it today, he comes from a family of "freedom fighters": "people that have been escaping something." His mother, Ilona, and father, Steven, fled Communist Hungary in the middle of the night in 1968. They were seeking a better life in America— and they found it. The price was leaving behind their infant daughter, Gabriella. The people-smugglers wouldn't transport infants. The Gyurkos ran a successful plastics company in California that provided well for them, for many years. And their son, Zoltan, was born on US soil (South Bay Hospital, Redondo Beach) in 1973. Gabi, however, the child left behind— though she joined her parents in America after three-and-a-half years—never truly got over her parents' original sin. They had fled Communist Hungary for material gain—for the chance to make more money and enjoy more comforts, realize their potential. They chose that over being with their daughter no matter what, even in a poor and oppressed country. "I don't think my parents and I ever recovered from that," Gabi told me.[1] She moved out of home the day she turned 18. Zoltan had the same reckless ambition that had driven his parents from their homeland. To put his life on the line in pursuit of greatness, damn the consequences... Heck, at age 8 he used to stare at the sun, thinking that if he just stared at it long enough, he would always be able to look at the sun without blinking or squinting. He had that recklessness and bravado, that determination and will, from a young age.

All life is lived on the edge, if you think about it. We're all just one accident away from the abyss, or too few too-short years from eternal nothing. Zoltan's choices had put him closer to the abyss, more often, than most. He knew death was present in moments like this, as black descended from horizon to horizon, and the elements screeched and lashed.

The storm was raging now. Winds nearly at hurricane speeds.

A sheer wave slammed his boat—a mast wire snapped. He ran from the cabin, and the wire slashed his cheek. His hand went to his face. "Fuck!" The cut was deep.

He tied up the wire, then tried to wait out the storm in the cabin; tried to read. All the elements howled and hissed and spat, just beyond the "safety" of his rocking box. *The Way* was "thrown around like a plastic bath toy," he later wrote.[2]

For two days, there was no let-up in the 35-foot waves. Water rushed over and through the boat and shorted his batteries—he couldn't use his electronics.

His cheek was infected and pus oozed down his face. He'd tried to sew up the cut, but the sea was so violent, he almost jammed the needle in his eye.

On the third day, he stood on the deck, holding on for his life... *When would it end?*

You could see existential risk coming on the ocean. The wave that could take your boat—and your life—out. As the sea roiled and the winds wailed around him, he saw it, building in the distance; growing from the surface—like watching a wall of skyscrapers rise in rapid time: stories-high tons of water that would collapse on him and there was nowhere to run.

He breathed in, filled his lungs. He shut his eyes and mouth, tight—he knew he was going under. He tied himself to a rope wrapped around the mast.

Then the wall of water fell and forced him under, upside down and trapped by the pressure.

He couldn't breathe.

He couldn't see.

He couldn't hear.

There was just water and blackness and one thought...

Could he survive?

***

"We will be the sun!"

I remember this phrase from the night I met Zoltan Istvan Gyurko (pen name: Zoltan Istvan) in person. We were upstairs in Carluccio's restaurant on Dawson Street in Dublin, Ireland. Up till recently, I'd known him only by reputation. He was a controversial Californian; a writer about the future of the human species, who supported genetic engineering, merging human minds with artificial intelligence, and replacing our limbs with robot parts—in order to lengthen our lives and make us better. It was a philosophy called "transhumanism." A set of ideas about overcoming human inabilities by transitioning into an upgraded species, transcending our natural biology through technology.

Zoltan was not the first, nor would he be the last, to have these ideas. But he had become one of their best-known advocates. I'd followed his run for president of the United States in 2016, which got worldwide media coverage. I'd written a book about transhumanism, which had just been published, and covered Zoltan's work in detail. That was the reason for this dinner. He'd come to Ireland, my home, to take part in a debate at Dublin's Festival of Politics. He'd been kind enough to say a few words to launch my book, and now he, my wife Sandra, and I were devouring pasta and wine.

I'd asked Zoltan what was the point of trying to live forever through technology, when we know that, in five billion years, the sun will become a red giant. Life on Earth will be impossible.

Life on Earth, perhaps, Zoltan suggests, but we will have left our solar system by that time, and we'll be a form of life completely different from what we know today. Maybe living stars. "We will be the sun!" he laughs, arms outstretched in mock triumph.

I'm skeptical that we'll get there. Sandra doesn't even like to eat GMO corn.

Later that evening, Sandra got a bus home, while Zoltan and I downed a couple of pints of Guinness (him) and a couple of Jameson whiskies (me) in The Palace, one of the most traditional of Dublin's traditional pubs. It was, as usual on a Saturday night, packed, but we got lucky and had a tiny table to ourselves upstairs.

Zoltan's ascent to his current notoriety had begun in 2013, when he'd self-published his novel about the future of the human race, *The Transhumanist Wager*, which became an internet sensation. When I'd read it, I concluded that he was a lunatic, and possibly a dangerous one. That may still be the case. But, having got to know him a little over Skype in the past few months, and especially now chatting with him over drinks in the bar, I had to face the fact that he was also a lovely man.

A person full of joy and ideas, who had lived—and was still living—an incredible life.

He'd sailed around the world for seven years in his twenties. He visited six continents, went diving for treasure, and was almost boarded by pirates. (He kept the pirates at bay by offering them a carton of Marlboros, a bottle of Sri Lankan whiskey, and $50—which was all he had aboard.) He'd worked for *National Geographic* in war zones, then returned home to the US and founded a business that made his fortune. Now he was devoting his life to the cause of living forever through science— writing about it, speaking about it, and running for office. When he put forward controversial or seemingly crazy points, it wasn't coming from a place of ignorance about the world or about life. It was coming from his lived experience.

His transhumanism advocacy had grabbed a lot of attention. The *New York Times Magazine* did a major feature on his 2016 campaign; writer Mark O'Connell traveling with him on his "Immortality Bus" tour across America. He traveled cross-country trying to convince people of the merits of living forever. And the *NYT* was just one of many outlets, national and international, that did big articles. He'd been interviewed by podcast and YouTube kings like Joe Rogan and Dave Rubin. He'd been bashed by Fox News, criticized on Vatican Radio. He'd debated oddball multimillionaire John McAfee. He'd spoken at the World Bank. The director of the Nobel Prize Foundation had even listened to him speak. The US Navy had asked for his advice about brain implants. *Playboy* had said: "When we're all glowing, immortal orbs, we'll probably have Zoltan Istvan to thank."[3] He'd founded what he called the first science-focused political party in history, the US Transhumanist Party, attracting followers around the globe—and making him the only 2016 US presidential candidate to be interviewed by hackers' group Anonymous. He'd been called an agent of the antichrist, and worse.

Basically, Zoltan was the kind of guy whose story you wanted to hear.

A lot of his antics had annoyed the hell out of transhumanists themselves. The movement grew out of technology-obsessed discussion groups in the 1980s and 1990s, finding its wings as the internet took off. In the second decade of the twenty-first century, it had gained some establishment credibility. Academics and major publishers had started to take its ideas seriously. Old-school transhumanists who'd been part of the movement for decades worked hard to counter the notion that it was somehow extreme or a joke. Zoltan brought a bomb to the tea party. He reveled in stunts like traveling around in a coffin-shaped bus; sticky-taping his bizarre manifesto to the US Capitol building; or infiltrating a megachurch to hand out atheist pamphlets. Other transhumanist campaigners promoted the philosophical roots of transhumanism in humanism itself; the idea that we should always better ourselves. Or they advocated for incremental improvements and the rigorous scientific method that had given us vaccines and cochlear implants. Zoltan wrote about requiring parents to get a license to have children, or artificial wombs solving the abortion debate, or bringing everyone who'd ever lived back to life one day through highly advanced 3D printing.

Zoltan insisted that transhumanism should be both a popular and a radical movement, not an academic talking-shop.

Zoltan's was a story you wanted to hear. Even if you didn't agree with what he had to say.

That night in Dublin, we talked transhumanism past and future, his battles with the old guard, our favorite writers and movies...

"Let's look at a video of my kid real quick!" he interjected at one point, having just been sent a clip of one of their daughters by his wife, Lisa.

His ability to go from the farthest ends of the future to the mundane family present, in the space of a sentence, is one of the things I enjoyed about chatting with him.

I would have stayed out for a few more drinks in The Palace, but Zoltan insisted on heading back to his hotel after two. He was traveling across the country tomorrow to go surfing.

I already knew by that point that we would have many more conversations—albeit over Skype.

A few months before, Zoltan had come across me online, and read about my book. He suggested we collaborate on something. Well, he suggested that I write a book about him. Now that we'd met face to face and we got on well, I was excited to get started.

I can't help thinking there was a certain cynicism in the reason Zoltan approached me to author his story. He'd wanted for a long time to pen his own memoir—he obviously had enough of a life story behind him—but he couldn't find the time with all his other engagements. He could have approached any of the far more accomplished writers who'd traveled with him on his Immortality Bus, who'd covered his campaign; any of the science and tech correspondents and feature writers and editors he'd worked with in his transhumanist life and his own prior career as a journalist...

But Zoltan knew that my own interest in transhumanism was not casual—it was personal. I was born with a rare disease, Fanconi anemia (FA), which causes the bone marrow to fail, so you can't produce new blood cells to keep you alive. A bone marrow transplant can give you a fresh shot at life, but it doesn't "cure" FA. Even if a transplant is successful, the patient is still at huge risk of developing head and neck cancers and other life-threatening ailments. The average life-expectancy for a Fanconi patient is pretty low. When I wrote my memoir about my bone marrow transplant, published in 2008, life-expectancy was 22. It's increased by more than a decade since then, and continues to rise as genetic and medical science advance. But people continue to die far too young. The night Sandra—who also has FA—and I met, I was giving a talk at a fundraiser in Canada. We sat at a table with two other 20-something-year-olds who had FA, and the next day I went to Niagara Falls with the fundraiser's organizers and their son, who had FA. Less than seven years later, as I write this, Sandra and I are the only FA sufferers from that group still alive. Sandra has lost 13 friends to this disease. I've personally never cared about living forever. But to live at all... To live as long as I have (I was born in 1984)... To live anything approaching a "normal" life-expectancy... I've had to depend, and will depend in the future, on advanced science and genetic research, on medical technology. The fields that could also one day deliver lifespans far beyond what we today consider "normal." Like Zoltan—albeit in a very different way—I confronted death early and had an interest in staying on the other side of it for as long as I could. When he came across my story online, I guess he thought this gave me a perspective that others don't have.

We also share an affinity for another, long-dead writer—which is not a trivial fact in the telling of Zoltan's story. But more on that later...

This isn't a ghostwritten autobiography. It comes from my interest in Zoltan's life and in his activism, but we disagree on much, and I've tried to be objective. He hasn't paid me a dime. The book is in many ways a conversation between me and Zoltan. (Indeed, whenever you see a quote that isn't attributed with a footnote, that's a quote from my many Skype interviews with Zoltan, between 2018 and 2020.)

In Zoltan's view, the stakes for this moment in scientific time couldn't be higher. We have the power to end all death—the power to live forever. Almost. We just need to keep pushing till we get it. Every 15 years that pass, another billion people are lost... Zoltan sees his activism as a battle to save all those lives. I don't know if that's a battle we can ever win. But I'm willing to hear him out.

Chapter 2

Sailing to Selfhood

He was seasick for ten days straight. Zoltan had never spent a single night on the open ocean, before he set sail from LA. He would now spend 20 days and nights, alone and far from land, before he reached Oahu, Hawaii. He had no engine. This was before GPS was widely available. He had to look to the stars to navigate. And he didn't know how to use his sextant—he'd have to learn fast.

It was August 1994, and Zoltan was 21. He'd been a champion swimmer in school, a nomad surfer in Costa Rica, a jailbird, and a student at multiple colleges. He'd been high—a lot— and been busted as a teenager for selling drugs. He probably already had more life experience than the average blond white guy from California. But when he set sail on his attempted circumnavigation, it was only then, he says, that his real life began.

He didn't tell his parents the truth, or give them a proper goodbye. He'd told them he was sailing to the Channel Islands— the "Galápagos of North America"—only 30 miles from Santa Barbara, where they lived. He didn't mention that there was no final destination, that he didn't know when he'd be back, or when he'd see them again. They would only worry if they heard all that. Worry about what had become of their "good son," who'd been tagged as gifted in first grade, and who had once diligently attended Catholic school, reading his Bible and going to mass; their proud Junior Lifeguard, their national swim-champ, whose life they had laid out for him: He would make the US Olympic swim team, hopefully win a gold medal, and then return home to take over the family business.

Zoltan didn't want Steven and Ilona to worry. But the truth was, by the time he set sail, he had already left his parents behind, in so many respects. Ever since he'd caught his mother's wrist, at age 12, when she'd tried to hit him with the wooden spoon one too many times. He'd rebelled against the strict upbringing of his immigrant parents. They'd pushed him to swim competitively from the age of 5, as soon as he'd shown talent in the water, training him hard—to fulfill his father's dream of having a champion athlete as a son. (Steven Gyurko had wanted to become a professional wrestler back in Hungary, but the hardness of his life and the need to put food on the table nixed that dream. His son was his avatar now.) Zoltan quickly grew comfortable in the water, comfortable with the solitude of the training and being alone with his thoughts. It also gave him a sense of discipline that wouldn't quit. But as he became a teenager, he'd had enough of competitive swimming. He ran away to Costa Rica, to pursue his ambition to be a pro surfer. His parents hated the idea—surfing was for hippies. But by the time his disapproving parents had realized he was actually serious about the surfing thing—that he didn't want to swim competitively anymore—and was actually intent on going to Costa Rica, 16-year-old Zoltan was already there. (He had a passport, and that's all he'd needed to get on a flight.)

Costa Rica was what cemented the adventurer in him. The coast of the Central American country was a blank slate for adventurers, undeveloped and lush. Expats and vagabonds—particularly from North America—were coming here to soak themselves in nature, and live free of the rat race. Zoltan began writing regularly, for the first time, keeping a journal. Inspiration was endless. Costa Rica was unconscionably beautiful—emerald and ivory and sapphire; rich jungle giving way to silk beaches... and, of course, the waves. For two months, teenage Zoltan hitch-hiked along the Pacific coast and rode some of the most celebrated waves in the world.

Wandering around high in the Costa Rican wilderness, he also discovered his mind. There was more to life than the physical ambitions he'd been pursuing: Be a champion swimmer, become a professional surfer. There was this whole depth of untapped meaning inside his head. It wasn't long before he started reading books, really reading for the first time—just to explore the lives of others, and the meanings in the world. He quips: "Once you start reading books, it's all over."

True to the cliché, it was at university that his intellectual life truly began. At San Diego State University, Zoltan started voraciously reading books. He names three he read in his first month at college, three American greats from two different eras: *Walden*, Henry David Thoreau's classic 1854 account of living in a cabin in the woods; *Leaves of Grass*, Walt Whitman's seminal poetry collection, the first edition of which was published a year later; and *No One Here Gets Out Alive*, Jerry Hopkins and Danny Sugarman's 1980 biography of the Doors' frontman, Jim Morrison. A manifesto for independence; an elaborate reflection on human nature and the human as part of nature; and an account of an iconoclastic genius who became the face of a counterculture. These three books, Zoltan says, destroyed the thoughts he had of becoming a professional surfer. He wanted to write. He wanted to make a difference through his thoughts and his mind, through intellectual pursuits. He wanted to pioneer in that way.

And indeed, it was a book that had inspired his latest adventure: to sail around the world, gathering all the knowledge about the world that he could—*Dove*, a bestseller in the 1970s. Written by Robin Lee Graham and Derek Gill, it charts Graham's round-the-globe sailing trip, which he began in the summer of 1965 when he was 16. As soon as Zoltan read that "wonderful" book, he knew he had to go sailing. It would be the ultimate freedom. He'd fill his boat with books and just travel, read, surf, and fish. He wouldn't be dependent on anything except the wind and his wits.

He earned enough money bagging groceries and as a pizza-place doughboy to buy a 26-foot, 1966 Pearson Commander sloop, which he named *The Way*. It cost him $2,700.

Zoltan spent another five weeks working on the sloop, learning its quirks, toughening it for the waves to come. He used to fix Enduro motorcycles with his dad. This was a different kind of job, of course, but that work had made him comfortable with fixer-uppers and with tools. He'd also raced those motorcycles—the rush of wind flying past. He couldn't wait to experience the new thrills of the ocean wind...

The sloop was a tiny vessel. No toilet. He couldn't even stand up straight in the cabin. He customized it for his needs, down to custom-built bookshelves (he made himself) to host his personal library, behind his bed in the cabin. Scouting out used-book stores and libraries, Zoltan gathered volumes in every genre to take with him. From Samuel Beckett to Ayn Rand, naturalist poetry to biographies of Alexander the Great and Hitler... From histories of nations to self-help bestsellers like *How to Win Friends and Influence People*... Before long, Zoltan had 556 books, still waiting to be read, that he would take with him on *The Way*.

Zoltan set out on his attempted circumnavigation with typical reckless bravado. "Sailing's really easy, you actually don't need any experience," he says. He taught himself just by reading, observing, and trying. Being in the water was the most natural thing for him, and the boat was just an extension of his body.

He quickly discovered, however, that life on the open ocean was very different than returning to dock each night. The stakes were much higher. There was no escape from storms, mechanical failure—or seasickness.

The stories from this time could be their own book, or a movie. This was the voyage that made him who he became. He wrote on the first page of a journal he kept while he sailed: "And so here begins a journey, not one across oceans and seas, or to lands and peoples afar—but one unto the furthest reaches and the most untamed horizons of the heart."[1] The journey was across seas, and to lands and peoples afar, but what mattered was where it brought *him*.

The sloop *The Way* was named for a poem Zoltan wrote right after high school. He named his boat after his own poem, already showing the egoism that would push him forward throughout his career. The poem "The Way" is filled with Zoltan's early preoccupations—women, drugs, and hating Christianity (his mother was a strict Catholic). Women "Take you anywhere they please, / With ease." The Zoltan in the poem "fucks" the serpent from the Garden of Eden "in the mouth / Again and again and again."

Beyond the teenage romanticism, the core of the piece is a watery metaphor about seeking truth, an authentic self, in the depths. Giving up pretense—"acting" how society expects—and becoming real:

Take the swim home friend
Frigid waters of courage
Move to the somber bottom
Then I'll show you the way
I know the way

Forget your act,
Forget your act
The pool is deeper than that.[2]

In his writing—his journals, his poetry—Zoltan was struggling to lay an intellectual foundation for a life that would be uniquely *his*, not something his parents or anyone else wanted or expected. The boat trip was a way of making his internal search literal.

15

Zoltan had taken to heart a message that was at the core of his beloved *Dove*. Graham put words on the appeal of solo sailing so well in his memoir: "I knew...there was something 'out there' that I desperately wanted. It was a chance to be my own man, a conviction that I was born free." It was a rebellion against "people determined to arrange my life in tidy patterns, prodding me this way and that until I could be safely sent out into society."[3]

Leaving from California, Zoltan didn't have a plan, as such. The only plan was to sail and see where it took him. He thought he might sail west all the way to India, crash his boat on the reefs, and hike into the Himalayas. He'd read that *nirvana* was to be found high in the icy mountains... But there was a long way to go, and it was too early to say.

He didn't expect the first leg to last as long as it did. "I just set my automatic steering and kind of went west." He finally found his sea legs—and found Hawaii, too.

His life in Hawaii set the tone for the trip: beaches, surfing, hanging out. No routine that couldn't be changed, except that every night he returned to his floating apartment anchored offshore. From America's 50th state, he sailed onward into the Pacific. He went south as far as Fiji and Vanuatu, stopping at Kiribati and Samoa. Then he sailed north to the Solomon Islands, Papua New Guinea, and up to Guam, where he took a job as a treasure hunter.

You might have a vision in your mind of a lonely life on the open seas, stopping only to restock on supplies and then keep moving. Although there were long stretches sailing between destinations, most of Zoltan's "sailing trip" was spent on land. He'd anchor the boat, chill out on the beach, and with locals. Go off exploring. He'd often read seven or eight hours a day, making his way through the hundreds of books he'd brought. "The books are the gold on this cargo ship," he wrote in his journal.[4] He spent three years hopping among the islands of the South and North Pacific, and this was how his days were.

He funded his life with credit cards. He'd take out one with a limit of a few thousand dollars; when it was close to maxed out, he'd take out another one, and start paying off the first card with the second. He lived frugally—there was plenty of free food in the ocean, he just had to catch it. He'd stocked *The Way* with a year's worth of canned goods before he left, too. The "floating apartment" had been carefully provisioned and kitted out to make sure he had everything necessary to survive and thrive. But he still needed money for repair parts, extra provisions, visa and port fees, and whatever treats and luxuries (beers at a beach bar) he chose to enjoy.

"I just lived this great life," he tells me, "from the point of leaving Hawaii to the point of getting to Guam, where I experienced amazing things almost every day." At his first stopover after Hawaii, at Tabuaeran Island 800 miles south, he contracted hepatitis A. He was sick for six weeks with the liver virus and lost an enormous amount of weight. "All these great things," he says. "I mean, it sounds like it wasn't great, but it actually was a very interesting time to be quite sick and then quite alive."

He spent six months in total on Tabuaeran, chilling with the indigenous islanders. There were no cars, electricity, or other tourists. He was amazed to discover the islanders had never tasted Coca-Cola. He danced and went hunting with them and surfed every day.

Each experience, good or bad, was another entry on the list of things he could say he'd done, or that had happened to him, in his life—and this was what most excited him. Every day of the trip brought him closer to a life filled with every possible human experience.

Zoltan witnessed and was part of strange and remarkable events. "I remember, for example, in Western Samoa, somebody kicked a poor dog that was blind in one eye—kicked it in the head, in the blind eye." He almost fought the man who'd done it, because he was so angry at the cruelty. "These are the kinds of strange experiences I had."

He shared kava with tribal chiefs in Fiji—he presented kava root seven times to seven high chiefs at seven Fijian islands. Presenting the root was part of a welcoming ceremony granting him permission to anchor offshore.

He learned to climb a coconut tree, and got a SCUBA certification.

When he reached Espiritu Santo, Vanuatu's largest island, he heard about native tribes in the interior who lived entirely without modern conveniences. They had seen few outsiders, though they were open to visits, he was told. He spent two days trekking inland, through jungle and across rivers. Zoltan was greeted by the Mareki tribe's high chief. He was the first Westerner ever to meet the Mareki, staying four days with them.

The Mareki way of life was so far removed from the way of things in California. Their villages were built in jungle clearings, the surrounding trees and hills the only protection from frequent hurricanes and heavy rains. Their homes were huts expertly constructed from tree branches, vines, and thatch. There was no technology, aside from simple tools made from wood and stone. There was no medicine, except what could be found in nature. There was no money. Tribespeople plucked grapefruit and breadfruit from the jungle, raised chickens and pigs, hunted lizards, and collected river crabs. What few clothes people wore were fashioned from plants and tree bark. Everyone had an established role in keeping the community going. The young men hunted and the women did all the cooking; medicine men gathered remedies from the forest.

Zoltan was a source of amusement and entertainment. The kids wanted to know why his skin was white—what was wrong with him? They studied him like an anthropological subject. His flashlight was fascinating—how did it do that?

Years later, Zoltan revisited the Mareki, when he worked as a reporter for the National Geographic Channel. He filmed a series of vignettes. He's seen being welcomed by the same chief who greeted him seven years before. He shows a photo of himself with members of the tribe from his first visit, pointing out to each villager where they are in the picture. It's a source of more fascination.

Zoltan documents tribal life in the series, and lives that life himself, looking for shellfish in the river and shooting arrows at lizards. Learning the tribespeople's language and observing their customs, their songs and traditional dance. He marvels that all the villagers need is nature and each other.

He remarks at the end of one of the vignettes: "My days in Mareki go by at a pace that we in the West lost centuries ago. It takes a while, but eventually, it stops being a little strange. It becomes normal."

Zoltan is seen perched on a tall rock, wearing just shorts and hiking boots, surrounded by nothing but jungle and mist.

On the voiceover, he says: "I realize that I'm not missing the twenty-first century."[5]

***

Life at sea suited Zoltan. Early in *Dove*, the narrator laments the solitude of sailing alone. It's clear that loneliness was a searing pain to be overcome. Zoltan found it tough, too: to have no communication with anyone, not even by radio, for weeks at a time. But he was also better prepared for it than most. He'd been alone with his thoughts in the water almost since he could walk, as a swimming prodigy, and he liked living in his own head. *The Way* was his personal space, for reading and thinking, writing and exploring.

In the decades to follow, Zoltan came to view his years on the boat as a libertarian touchstone: how easy it is to be free; to escape your current life for a better one, if you really want to. He wrote on social media in 2020 that "even some of the poorest people in the world could afford" the lifestyle he had on the sailboat: "People say the American Dream is dead. Maybe because you can't reach it says something about you, and not about the system. To me, the American Dream is very easily reachable—but you can't have four children in your twenties, nor can you shop for every trinket at WalMart." Why not buy a $10K rural lot and build your own home with materials from Home Depot? "You can learn all the necessary skills easily off YouTube videos." And the materials might cost only $25,000.

"I just don't understand why people are so angry or poor," he wrote. Why aren't they buying boats to live on, "traveling all around or building their own homes and making equity"? Of course, not everyone can live rent-free in their parents' vacation house in Brookings, Oregon while they save for a boat. Not everyone gets off so leniently after a drug bust—with just a month's jail and 11 months of probation. Zoltan wasn't blind to the role of privilege and luck in his own success. But he couldn't stand the fact that, everywhere he looked in modern America, "healthy, able-bodied people" were filled with despair and a sense of futility.[6] He just didn't recognize that mindset. His whole mission from the time he reached his mid-thirties was to jolt folks into reimagining what was possible for them. To do that, he drew on his seminal years of freedom and self-discovery in his twenties aboard *The Way*.

When I suggest to Zoltan that not everyone has the advantages he had, he brings the conversation back to his mindset. His work ethic—the fact he was willing to put in 70- to 80-hour weeks from a young age. And his willingness to reframe obstacles as advantages (a drug bust isn't a problem, it's a great story you can tell about yourself). Say he did have to pay rent in Brookings, and couldn't rely on his parents? So what? Rent was a few hundred dollars a month, and paying it would only have slowed down his departure—not changed his direction. He paraphrases Hermann Hesse to me: If you've no food, the easiest way to get by is to fast.

Sailing was the life he wanted, but it wasn't a breeze. There was always stuff to be done, maintenance and fine-tuning to keep the boat in top shape and make sure he could stay afloat. Cleaning the air pollution (even in the middle of nowhere) off the deck and sides. Fishing for his next meal, then preparing and cooking it. Watching for swells and storms. Navigating and staying safe in the squalls when they struck. Not to mention simply finding his way from land mass to land mass, judging the stars and the moon and hoping he didn't miscalculate.

He ran aground seriously only once in the seven years. It happened as he headed for Gaua, one of the Banks Islands, at the north end of the Vanuatu archipelago. The 130-square-mile island rises into the active Mount Gharat volcano. Zoltan was snapped awake at 6 a.m. by a strange sound—*The Way* was crunching against coral. He sped to the deck and looked up to see he was fast approaching Gaua's cliffs—a crash that would cripple his boat. He wasn't supposed to be in this area at all. His navigation had been off by 7 miles. He frantically tried to steer the boat against the heavy surf, dropping his sails to slow his momentum. Then he dropped his anchor.

The sloop swung around. At least he was no longer going to hit the cliffs. But *The Way* was now lodged against the coral. The water was only 3 feet deep. Zoltan revved up his outboard motor, but the vessel wouldn't budge.

Then he had an idea. He started tossing supplies and equipment overboard—dozens of cans of food, then his three 5-gallon water tanks, his dive belt, even the engine for his dinghy.

He'd made the sloop a few hundred pounds lighter and, thus, easier to maneuver. But there was still one more thing that had to go overboard before he could move his boat. He threw himself into the water. With the outboard motor running, he pushed the boat out from the coral by hand. He half-waded, half-swam back to the open ocean, guiding *The Way* alongside him.

When he inspected the damage, he saw that the underside was battered but there were no holes, so there would be no leaks. His rudder was destroyed. He'd have to rebuild it, and replace the lost food and equipment, at the next port.

Zoltan wrote in an article about the grounding for *Ocean Navigator* magazine: "Sleeping was never the same again. For many years after, I awoke in the middle of the night, panicking, running out to the cockpit, ready to steer the boat through surf."[7]

One bad day at sea cast a long shadow. But there were far more good days than bad. For every challenging moment—a storm, a venomous snake that got aboard—there were ten moments of sheer bliss. Peacefully reading with the waves lulling him. Watching schools of fish outpace *The Way*. Being the only human as far as the eye could see. Another dawn, another dusk to watch from his deck; the star that gave life to the Earth rising and sinking for him alone.

There were times when even he felt lonely, when he felt the weight of his solo undertaking, and thought about the conveniences back home he'd given up. But there was always another island to visit, another stopover where he could enjoy months of company and fun, another wonder waiting to be uncovered, just over the horizon. Another "out there" adventure or culture shock—like climbing active volcanoes, which he did in Papua New Guinea, where he also dropped in on the Asaro tribe, the highland "mud men" famous for their ghost-like masks.

By the time he got to Guam, Zoltan had three credit cards on the go, all paying for each other, while they also funded his life. He needed a job and real income to start paying down the debt. He began working in a dive shop, filling scuba tanks, and was introduced there to another opportunity. He would take odd jobs occasionally through the years to come, working construction and saving cash till he could sail onward. But no job matched this one. He became a professional treasure hunter, an "archaeological salvage diver." A project had been ongoing on the south side of Guam since 1991, and Zoltan became a part of it.

From the sixteenth to the nineteenth century, Spanish galleons had traversed the Pacific, carrying cargo from Latin America to the Philippines, from one edge of the Spanish Empire to the other. In those centuries, over 40 of the colonial ships were lost at sea.

In 1690, a galleon had hit Cocos Reef off the coast of Guam, and sunk. It was on its way to Manila in the Philippines (and the Asian trade routes beyond) from the port at Acapulco in Mexico. The galleon, *Nuestra Señora del Pilar de Saragoza y Santiago* ("Our Lady of the Pillar of Zaragossa and Saint James"), had a crew of 120, and there were soldiers and missionaries aboard as well.

Everyone who was on the ship survived. Nearly all of the cargo, however, was lost. The *Pilar* was believed to be carrying up to two million minted silver pieces of eight — still waiting to be found, 300 years later, somewhere on the ocean floor.

It was an elusive prize... Amid the endless terrain of the seabed, the "Pilar Project" team hoped to discover the main body of the wreck and the "motherload" of silver coins — worth about one billion dollars in current money.

It was high stakes and high drama. Zoltan was one of a dozen divers who made the daily plunge to the ocean floor. The pay was good ($14 an hour), but they were putting their lives on the line for the paymasters. This wasn't regular diving with fins. They wore heavy belts, and boots, and plodded along the seafloor, mapping territory with ropes, followed by a videographer, who would record what they found. They had tools to break coral, suck away sand; metal detectors. During Zoltan's time, one diver was airlifted away with the bends. He heard that two years before, a diver had drowned.

It was a big adjustment. Zoltan had been captain of his own boat for quite some time. Now he had to take orders from an asshole of a boss who reported to far-off rich investors. He was stuck on a small barge with other 20-something-year-old guys, or combing the seabed with them, and they were all in competition—sometimes heated—to find the motherload. And it was the kind of find that, when you discovered it, "within hours you would have people with machine guns out there protecting it."

The young divers did like to dream. They imagined sticking it to the bosses. If they found the motherload, why even tell? They could just come back in the middle of the night and load up Zoltan's boat. Even if they snatched a small fraction of the motherload, they could come away with treasure worth 50 million dollars. They didn't even plan to keep it. The boys imagined they could take their 50 million and live among the Pacific islanders, share it with them and improve the islanders' lives.

Zoltan wrote in an article about his sailing trip in 2001: "Unfortunately, by the time I earned enough money to get sailing again, our team of divers still had not recovered anything more significant than cannon balls, pottery shards and 33 silver coins."[8] He found one of the cannon balls, worth maybe $1000 to a collector.

The journey onward to *nirvana* couldn't be delayed. He was glad of the chance to be a treasure hunter. He was also glad, after four months, to leave that volatile dynamic behind... To have more time again to himself, and for intellectual exploration. He hadn't even been allowed to read his books on the barge during downtime.

As well as his reading, Zoltan was still writing a lot. In his journal he recorded reflections and scribbled poems based on his inner life and outer events. The entries are heavy on introspection, and the workings of a mind trying to untangle the knotted ball of wool that is human experience of the world. In his thoughts in these years the tension between the human world and nature was paramount. His very first entry in his sailing journal, in 1994, set the tone for his thoughts over the next couple of years:

The Earth was once driven by a heart so engulfed by madness, a heart so vast and untamed—but one is hardly aware of that heart anymore. Its beat, once rampant and boisterous, is now quiet and soft. Man has tainted it—stowed it away and deemed it obsolete. He has manufactured his own heart—fashioned after safety and convenience—I should not even call it a heart. Where is its love? Where is its wildness? Where is its madness? It is a slow and rusting machine now—and nothing more...

...I hated the world—that world that modern man had shaped; it was rotting like some prized piece of flesh pillaged by maggots. But the heart was still there—albeit hidden well and unnoticed by the herd of men—I could still feel it—barely pulsating. And it drew me in. Oh the heart of the ancient world, yes—the heart of the universe, sweet fabric of existence itself... It was still there—but you had to cross—you could not remain protected and comfortable in the world—and still know it. It would be impossible then.[9]

The young Zoltan sought to connect with something eternal by turning away from modern society—the need for a steady job, accumulating stuff, convenience as an end in itself—and pursuing what he thought was a more authentic existence: striving against the elements, forced to be alone with his thoughts, not answerable to an agenda. His sail trip was, in a sense, a twentieth-century version of Henry David Thoreau's life in the woods, as recounted in *Walden*. Both *Walden* and *The Way* were an experiment in solitude and self-reliance, amid pure nature.

It's hard to reconcile the 21-year-old Zoltan who wrote the above—decrying the modern world and appealing to the essence of the universe—with the Zoltan who would write, in 2019: "Nature Isn't Sacred and We Should Replace It."[10] But look closer, and his condemnation of the modern world has nothing to do with its technology, and everything to do with conventional social expectations. Zoltan, ultimately, would find a salve for his youthful concerns in a passionate, wild, zealous, dangerous—individualist and libertarian—pursuit of his own agenda. And it was early on his sail trip that he met the man who became his greatest inspiration.

Since the start of the voyage—in fact, since long before, though the process became more urgent on his travels—Zoltan had been piecing together an idea of who he wanted to be. Like many of us who desire to improve who we are, he began to develop a character in his mind: the ideal version of himself.

Quite apart from setting goals for achievements he could attain or jobs he could do, this was about, as I say, his character. The ideal character he created might not be someone he always lived up to, or could live up to. But it would be a beacon. It would show him the way. He scribbled in his journal sophomoric thoughts about who this person could be. He hadn't found an ideal to live up to in the Bible (which he'd read from cover to cover in jail, and found to be absurd), or in his parents, despite his love and admiration for them. Maybe Jim Morrison came closest. But Zoltan didn't want to end up like the Doors' singer— substance-dependent and dying young.

Then, in Fiji, Zoltan met his ideal man. Or as close to a version of his ideal as it was possible to find. The man was tall and lean, with blond-red hair. He smirked a lot, as if the world of other people was something he didn't quite get, and it amused him. He was an intense loner. But Zoltan was drawn to him. The man liked to swim, alone. He'd been expelled from architecture school in Stanton, Massachusetts, because he'd refused to do the assignments the way the college wanted—he was only interested in drawing his own kind of buildings, angular-shaped things like no one had seen before.

The man had gone on to become celebrated and infamous. Undeniably a genius architect of unique vision—often compared to modernist Frank Lloyd Wright—he was also uncompromising to the point of extremes, believing that his own talent was infallible. He shocked America when he unilaterally dynamited and destroyed a public housing project he had designed, because his original plans had not been followed. He was put on trial for that incident.

The man had argued in court that artistic integrity was a higher law. That no one had the right to take it from him, and so he had acted in self-defense. He was acquitted, and took a commission to build the tallest skyscraper in New York.

Zoltan found the man and his story incredible—an inspiration from the first moment. Truly, these were personal qualities and a life trajectory he could hope to emulate. The man's strength of commitment to his own cause, prepared to suffer being outcast; prepared to endure any hardship, in order to stay true to his own mind and soul. This guy had bet everything on himself—and won big.

The man's name was Howard Roark. And he was the fictional hero of Ayn Rand's novel *The Fountainhead*.

Chapter 3

The War on Death

"What's the most important incident in my life? I would say, reading *The Fountainhead* the second time in Fiji is the most influential eight or nine days of my life. There's no question about that," Zoltan says. He finished the novel once, then opened it up the next day and started again. *The Fountainhead* wasn't even one of the books Zoltan had selected himself for his sail trip. A friend had given him a copy.

Zoltan's claim about his own growth is a bold one. He was telling me, in essence, that a book that was a half-century old by the time he read it had more of an effect on him than everything he had done and experienced up to that point, and everything he would do and experience afterward. More of an effect than running away to Costa Rica at age 16; more of an effect than surviving storms and pirates on a semi-circumnavigation of the globe; more of an effect than his travel to dozens of countries, including to war zones. More important than college—than the six colleges he attended—than meeting his wife, than the birth of his children. A book was this powerful to him.

The Fountainhead casts a spell on many. Its author, Ayn Rand, was born in St Petersburg during Russia's last years under the tsars. After Lenin's Communist revolution in 1917, the Red Army seized her father's pharmacy business. Ayn, only 12, "burned with indignation," writes Rand biographer Jennifer Burns: "The soldiers had come in boots, carrying guns, making clear that resistance would mean death. Yet they had spoken the language of fairness and equality, their goal to build a better society for all."[1] The forfeiture of the family's property, the deliberate destruction of middle-class life under the Bolsheviks, set Rand on a path where—fleeing to the United States—she became one of the twentieth century's most vocal and popular advocates of Communism's opposite: capitalism, and the rights of individuals to behave as they saw fit, free of interference from the masses.

Rand believed that she could make profound ideas into thrilling stories, and *The Fountainhead* (1943) was a breakthrough success. The book combines soap-opera intrigue (as various forces work to thwart Roark's career) with loaded scenes of passion (between Roark and his love interest, Dominique Francon). It takes us through the lows and the highs of Roark's life, ending on a note of triumph. Underneath all that, however, the tale of Howard Roark is a primer on achieving independence; from "the system," from other people's expectations, from tradition, from family—from whatever is holding you back. Not surprisingly, teenagers and young people, generation after generation, have been big fans.

But more than that—this 700-page novel helped shape the world. Another Rand biographer, Anne Heller, writes that it "almost single-handedly renewed popular interest in the cause of individualism" during a time when socialist ideals were ascendant around the globe.[2]

Rand expanded her defense of individualism in *The Fountainhead* into a full-fledged capitalist manifesto in her 1957 sci-fi opus, *Atlas Shrugged* (which Zoltan read a few years later). Fans of Rand's work—claiming her as a guiding light—became influential policymakers in the US government in the 1980s and beyond, including Alan Greenspan, longtime chairman of the Federal Reserve, and several lawmakers and members of the Reagan administration. Entrepreneurs and CEOs would cite her as an influence for decades to come, including PayPal co-founder Peter Thiel and Wikipedia co-founder Jimmy Wales. She also had a strong influence on popular fiction and within popular media, with fellow writers either inspired by her or referencing or critiquing her ideas. Spider-Man co-creator Steven Ditko was a major fan and even created stories and characters inspired by Rand's ideas, as did fantasy author Terry Goodkind.

Zoltan jotted in his journal after he finished the book: "*The Fountainhead* has stripped away the invisible monster I've fought my whole life—for the first time I no longer fought just a shadow—but a recognizable evil, a recognizable beast."[3] The "beast" was the second-hander, a type of person Rand names: someone who copies what society expects, doesn't think for themselves, and thus betrays themselves.

By the time he and I came into contact, Zoltan had read *The Fountainhead* 13 times. He often returned to it when contemplating the next steps in his life, how best to go about achieving his goals. He would ask: What would Howard Roark do?

In fact, the main reason Zoltan reached out to me to write his biography is because my book about transhumanism was also—indeed, primarily—about Ayn Rand: her influence on science-fiction writers and real-world technological fields, including transhumanism. Rand believed that the individual mind is the engine of progress. We must, each one of us, think for ourselves. The entrepreneur, the artist, or the philosopher with the best ideas is the one who can drive civilization forward. She celebrated creations and inventions that make life easier, or that make new things possible: plastics, automobiles, lightbulbs, skyscrapers, air travel, computers, moon rockets. She loved the conveniences and the opportunities of modern capitalism. (She would have loved Zoltan's parents' story; it's not unlike her own. How they fled Hungary, worked hard, and created wealth for themselves under capitalism in the US, and so were able to enjoy a standard of life that would never have been possible under Communism.) For Rand, capitalism is a creative arena where free-thinking individuals can advance their own lives as well as push the limits of human possibility. Because of this, she inspired entrepreneurs and inventors in Silicon Valley and early transhumanists.

Zoltan didn't know any of the history of Rand's influence when he closed *The Fountainhead*. He didn't know he'd just joined this history. He just knew his life had changed suddenly, in ways that would play out for years to come.

***

Zoltan had been reading and sailing in the Pacific for nearly two years by the time he took the archaeological salvage job in Guam. While he was there, he attended the University of Guam, collecting a few more college credits. His reading while on *The Way* had made him long for a more vibrant intellectual life. He was reading and writing all the time. He was developing his own concrete ideas about the world, and he wanted to discuss ideas with other people. He also wanted a decent college degree. As he wrote more and read more, he figured it would give him credibility.

When it comes to conferring credibility, in the eyes of some, nowhere is better than an Ivy League school. In Guam, Zoltan applied to two, promising to play water polo if he got in—a sport at which he excelled. He'd played on both his high-school team and with the University of Southern California Trojans, with whom he won a national 18-and-under championship. Now, he applied to Harvard University and Columbia. He was accepted into Columbia. He says he didn't really even understand what it meant—what an achievement it was to get into an Ivy League school—until his Hungarian relatives went crazy about it. His life certainly now regained some credibility in the eyes of his parents.

Zoltan anchored the boat in Guam, and moved to Harlem to attend New York's premier university. Big City Living—his first real experience of it—was whiplash after his boat and island life. He describes the building where he rented an apartment as a "drug den" and a "prostitute headquarters." He paid by the week. There was a barbershop on the ground floor, with apartments above. The floor below his was home to an apparent brothel. Folks smoked crack in the hallways; and when he looked out his window, there'd be folks selling sex and drugs. More than once he was trying to sleep and heard gunshots from the street. It was surreal, attending this stuffy school for the privileged while living where he did, right next to it. Columbia was a lawned, classically architectured oasis, with a revered higher-learning history; yet Malcolm X Boulevard, where he laid his head, had an even more revered history in many ways, as the locus of the Harlem Renaissance.

NYC hit Zoltan with sticker shock. Life on the boat had been rent free. There were no utility bills, and a lot of his food was caught for free, too. He'd been spending only about $100 a month buying produce and whatnot at island markets, and keeping up with repairs. Now he was in one of the most expensive cities on the planet. He got a job as a guard, manning a security desk at Columbia at night. This was perfect: He could study at his security desk, while he went to classes during the day.

The Big Apple was there to be plucked... The greatest city in the world, some said, including his literary hero, Ayn Rand—who adored NYC as the peak of civilization, a beacon of finance and entrepreneurship, industry and intellect. Zoltan wanted to take full advantage of the opportunities he had in Manhattan that he wouldn't have anywhere else. He spent a lot of time at the museums, the Guggenheim and the Met.

The schooling he found to be pretty mediocre. He transferred into Columbia with two years' worth of credits, between San Diego State, a brief stint at Los Angeles Harbor College, then the College of Southern Idaho, College of the Redwoods in northern California, and the University of Guam. And so, he would be able to emerge with a degree from an Ivy League university in a short timeframe. The biggest difference he noticed between Columbia and the other colleges he'd attended was the cost. Community colleges had cost him a few hundred dollars per credit. At Columbia he took on $80,000 in debt, for a year-and-a-half's work.

It was hard work, though. During his first semester he was brought up to speed on a lot of the basics, immersed in Columbia's famed "Core Curriculum," which all undergraduates take before their specialized major, and which includes Masterpieces of Western Art and Masterpieces of Western Music modules. Zoltan also studied college composition and elementary Spanish. He enjoyed his exposure to literature and art and music, but between his coursework, his night job, and the stress of where he lived, he often yearned for *The Way*.

He was an outsider of sorts at the Ivy League university. His life experience at this point was entirely different from his peers'. Maybe it was just his perception, but he felt the other students—most of them younger—looked down on him because he hadn't followed the usual route to Columbia. The friendships and the social life that he had were mostly based around water polo. "All the wonder and all the experience I had didn't translate very well to anybody," he says. "Most people just thought I was a weirdo."

He did appreciate all the drugs available to him. One of his best memories is one particular night. He took part in "The Odyssey," an endurance test set up by the coed Greek house on campus, Alpha Delta Phi. This was in his early days at the university—and quite an introduction to New York. He took part in the hazing with eight others, most of whom wanted to join the fraternity. Zoltan was just there for the thrills.

The journey began in a tent at the fraternity's brownstone. First order of business: Drink the mushroom tea. Then Zoltan and the others were brought on a midday walk through Harlem. Everyone was tripping.

Suddenly, a van pulled up—the side door sprang open, and people with machine guns began demanding they all get into the vehicle.

The tea-drinkers were bundled inside, and the van took off. One of Zoltan's companions was crying. "No, no, it's OK," someone brandishing a fake machine gun reassured them. "This is part of The Odyssey."

There followed a 24-hour "drug-a-thon," where they were taken to different locations around NYC, and fed different drugs. There was booze, weed, acid... Every hour there was a new location. They went to the Aqueduct Racetrack in Queens, and bet on horses, high. They inhaled laughing gas on Coney Island. They went to the Jehovah's Witnesses' headquarters in Brooklyn Heights, wearing purple wigs, and asked for paperwork to join up. The actors of the fraternity put on a play, high, in an abandoned Brooklyn theater. At an empty house, everyone looked through a window while white-coated men behind the glass mock-electrocuted another student. They danced in Times Square, wearing the same purple wigs. They showed up at the famous Webster Hall nightclub. They ended The Odyssey at New York's Russian Baths, immersed in rejuvenating waters and sipping chilled orange juice.

So much about those 24 hours made Zoltan feel lucky to have had this chance... And made him crave even more unique experiences, and even more from his life.

***

When did you stop thinking you would live forever? As kids, we have no concept of death. The world is eternal, and our life simply goes on. We're not equipped to think too much about where it began and what happens in the future.

At some point, we realize that some people are no longer here. Their experiences of this world have come to an end. Grandparents; a friend caught in a tragedy. But we're told, most of us, that they still exist somewhere else: There's another realm where we live on in another form.

I grew up in a Catholic country. Most people around me believed in an afterlife. Both my parents were skeptical of religion, and hated the power the Church historically held in our country. So I was raised agnostic or atheist, some cross between the two. Raised without any religious instruction, one way or the other. I was baptized Catholic to please my mom's mom; my dad refused to go to the ceremony.

I didn't know anything about heaven, but I knew more than the average child about death — or the specter of it. When you're diagnosed at age 9 with a killer illness, it hangs over you. My wife, who has the same condition, told me not long after we married that she had lost 13 friends to the disease in the two decades since she was diagnosed. Only war takes more bodies.

It's easy to imagine living forever, just going on and on. Most of us don't think too much about death until we see it approaching. Some of us think we'll keep on living — just somewhere else, in heaven or an afterlife — even after we depart the earth. But immortality is not just a concept for the religious. How does an atheist get to live forever? Well, without any belief in an afterlife, we have to make it happen for ourselves.

By the 1980s and '90s, a lot of atheists had started believing immortality could be possible, not through a religious savior but through technology and medicine. After all, life-expectancy had increased steadily for decades. Quality of life into old age was only getting better. Diseases that were once rampant killers, like smallpox, had been eradicated. Surely it was only a matter of time before we knew how to rejuvenate the body, so people could keep living indefinitely. Maybe we could even bring people back to life. As computers advanced, maybe we could transfer our thoughts into computers, and our minds could live in cyberspace or in robot bodies. It may sound like science fiction, but elements of science fiction often become some kind of science fact. Test-tube babies and IVF have been compared to Frankenstein's monster. Many now see the *Star*

Trek "communicator" as a precursor to the flip-phone. Video calls in *The Jetsons* became Skype and Zoom and FaceTime.

I first read about cryonics—freezing and preserving your body or head when you die, in the hope that you can be revived in the future—in a 1990s British paranormal magazine called *The X Factor*. I was just 11 years old—but once you read that such a thing might be possible, it's hard to forget. Especially when you're facing a life that might not be that long.

The article painted a picture:

You slowly open your eyes.

Your head feels slightly sore, your throat is dry and your whole body feels numb, cold. As your eyes begin to focus properly you see a man in what looks like a space suit peering down at you. He smiles.

"You've made it. Welcome to the 25th century."

...

The theory behind cryonics is temptingly straightforward. You die, you are frozen and then you thaw out in a future world where every conceivable technical advance has taken place. All your problems, including mortality, are solved at a stroke. Here, they can cure anything. One hundred or two hundred years after your death you get up and walk.[4]

Hope is a crazy thing.

Just a few months before I read those lines, Zoltan was discovering cryonics for the first time. He was older than me— it was during his first semester at Columbia—and it struck him with far more force. His English composition class was handed an article on the subject, to discuss and analyze.

It was a revelation: "There's a movement in America of people that want to overcome death," is how he describes his takeaway. There are folks out there trying to turn science fiction into scientific reality? Zoltan wanted to join them. Abolishing death, making it not matter, was beyond anything he'd imagined might be possible—and maybe it was impossible. But there was a movement working toward it. It was just the sort of giant challenge he'd been looking for, to give meaning to his life.

He mentally filed away what he'd learned, knowing he would return to it. "That was the moment that I said, 'OK, now I have a life goal,'" he says. This was a cause he could believe in. These folks weren't putting their faith in a god or the supernatural as a way to transcend the flesh. They were using science, the scientific method, to maybe one day eliminate death. Their methods were odd and unproven—but isn't that how some of the greatest scientific and technological achievements started out, as a challenge to the status quo?

By this time, transhumanism, and wider aspects of "out there" death-defying science, were receiving a fair amount of attention in the mainstream media. Whether it was cloning Dolly the Sheep or the internet seeming to create a whole new dimension of reality—a "cyberspace" to extend the mind—science and technology were fundamentally changing our relationship to life. Zoltan hadn't yet heard the word "transhumanism," didn't know it was a word he should be interested in. He just loved the idea of overcoming death through science.

Human beings have always imagined living forever. Most religions advertise themselves as a pathway to some kind of eternal life. A paradise awaits the soul after your physical body expires; a land that can't be seen on this side of the grave. Many commentators have compared transhumanism to a religion, just a modern way to fantasize about immortality. But for many transhumanists, unlike religious believers, transhumanism actually begins with the knowledge that there is nothing beyond the physical world, there is no spiritual realm awaiting us. There is no afterlife, no other life. If we want to prolong existence, the only way to do it is through science and technology. Lengthening your life on Earth is a very different prospect than getting into heaven because you just have faith.

In mythology, across cultures, there are stories of quests for the elixir of life, the secret formula for living forever; of people returning from the dead or challenging the life-giving power of the gods. Often, these are morality tales, warning us not to defy the natural order of things, not to go against the status quo, or we'll suffer a horrible fate like the characters in those stories: Icarus, Prometheus, Faust, Frankenstein. They're stories designed to make us content with our lot. And often they reinforce the idea that only faith in a particular set of ideas, from a particular religious authority, can ensure a better life in the beyond. But the existence of all these stories is a reminder that pushing against the limits of a natural lifespan is a human constant. It's as wired into what the human being is as anything else.

Transhumanists do not see their ideas as wacky, as weird science fiction or some kind of techno-cult. Rather, in the transhumanist way of thinking, one day, it will be no more weird to genetically engineer a baby without disease, or back up your consciousness to a computer, than it is to undergo an organ transplant, or take Prozac to alter your mood, today. Transhumanists see life-altering technologies like cryonics, mind-uploading, and genetic engineering as just more steps in the long line of medical advances that have enabled individuals to live longer and achieve more. In the last century, human life-expectancy has doubled, and that's thanks to how technology and medicine have advanced, with inventions that make life easier, and treatments that increase our years on Earth. If, 100 years ago, life-expectancy was 39, and today it is 81, why could it not be 160 later in this century, and then 320, and after that... Who knows? Continually renew and regenerate the body—keep aging always at bay? Live forever?

Progress is not inevitable. It requires constant work, painstaking innovation, like all breakthroughs of the past... thinking anew about what we can and should be as living beings on this earth. Transhumanists see themselves engaged in that work.

The word "transhumanism" was first used by biologist Julian Huxley in the 1950s. He used it to refer to new ways of thinking about the possibilities for human life, "trans" in Latin referring to "change" or "beyond." But the modern definition of transhumanism was developed by Max More, longtime president and CEO of cryonics facility Alcor, in the 1980s and 1990s. He wrote about transhumanism as a philosophy, a way of thinking where you consider modern technological and scientific life as a "transitional" state between human life as it has been and future post-human life, which we will create: artificial intelligence, genetically engineered beings, cyborgs. Transhumanism, as a belief system, sees this transition as a good thing, and wants to bring it about. More and his internet-based group, the "Extropians," did a lot to grow transhumanism, organizing events where ideas were discussed and developed, and publishing regularly. (The "Extropians" named themselves such, as they were advocates of "extropy"—a new term coined by More to mean the opposite of entropy. In other words, the opposite of decay and certain death: unlimited growth and perpetual life.) Eventually, the Extropy Institute melded into the World Transhumanist Association, another advocacy group founded by futurist philosophers Nick Bostrom and David Pearce, now known as Humanity+.

Like Zoltan, More and the Extropians were influenced by Ayn Rand, both her inspirational individualism (stand firm against what society wants you to be; do what you want with your own mind and body, even transform it with tech) and her argument that continual technological progress is good—in fact, it's one of the unique and best things about being human. (In *The Fountainhead*, Howard Roark tells us: "Every great new invention was denounced. The first motor was considered foolish. The airplane was considered impossible...But the men of unborrowed vision went ahead...The creator's concern is the conquest of nature."[5]) Around the same time Zoltan discovered Rand's ideas, they were being presented and debated at Extropian events and in their journals.

There are a lot of people associated with the transhumanism movement—by media commentators or the public—who don't necessarily like the term and don't necessarily refer to themselves as transhumanists. Folks developing artificial intelligences or working in CRISPR gene editing or inventing therapies that slow the body's aging process. But it's fair to say that these things are part of transhumanism, since they're about big changes that bring us beyond what humanity was previously capable of. That's the tension in transhumanism: It is, on the one hand, about a long tradition of human beings imagining longer lives and the slow, steady progress toward human betterment through the centuries; on the one hand, it's part of the normal tradition of progress in medicine and technology. If this is the case, why do we even need a word like "transhumanism"? Why not simply call it scientific and technological progress?

Well, because, on the other hand, transhumanism refers to radical change. While it may depend on legacies of innovation over centuries, no one can say that the prospect of robot limbs becoming normal; or revival after death—via cryonics—becoming normal; or living to 360 becoming normal, would not fundamentally alter what we think it means to be human, take us "beyond" what it is to be human right now. Transhumanism is the realm of exploring these radical, fundamental changes. Zoltan sometimes refers to transhumanism as the most extreme 10% of all science.

Transhumanism is first and foremost about the progress of science. But it does share something with religions, a kind of faith, the spiritual element in all human endeavor... Hope. The hope that we can be more; more than what we are now. That life in the future will be better. That death can be held off. A world without disease, where loved ones can be revived, where accidents aren't fatal... Only technology can make these things possible.

Zoltan read about the unproven technology of cryonics and it gave him hope... that there might be no limits to the life he could imagine. And hope that he could find his life's purpose— in helping to make real a world of unimaginable possibility.

Chapter 4

The War on Ourselves

The taxi hurtled through the streets of Kargil, in Indian-controlled Kashmir. Zoltan seemed to be the only foreigner in the city. He'd felt like all eyes were on him as he'd hurried to hail the cab. Now he was heading toward the fighting. The taxi would take him to the village of Chituk, near the Line of Control between Indian-administered and Pakistani-controlled Kashmir, where there'd reportedly just been a bomb attack.

It hadn't been easy to get this far. Refused a visa for Kashmir at the Indian consulate in San Francisco, he boarded a flight to Nepal. He was ultimately granted the visa at the Indian embassy in Kathmandu, "where the computers blackballing journalists had yet to arrive," he says.[1]

After graduating from Columbia in October 1998 with a BA in "Religion and Philosophy in Writing," Zoltan had taken to the seas again, back on *The Way*. This time he brought with him a girlfriend, Jennifer Hile, who'd been working as a magazine researcher for *National Geographic*. The couple spent more than a year adventuring. They took *The Way* up the Straits of Malacca, sailing off the coast of Malaysia, and moored at deserted islands off Thailand. After crossing the Indian Ocean, they traversed up the Red Sea, through the Suez Canal, out into the Mediterranean. They took time to visit Sudan, Egypt, Israel, Jordan... In all, they visited 14 countries across Asia, Africa, and Europe. Zoltan had now visited dozens.

After landing in Cyprus, Jennifer and Zoltan went their separate ways, but kept in touch. Jennifer got a job with the new National Geographic Channel when it launched in January 2001. She'd left her *Nat Geo* desk job to witness the world she'd only been researching... In her second stint, she wouldn't be desk-bound, but journeying all over to write, produce, and host fascinating stories about the planet, its wildlife and people. She helped Zoltan get a similar gig at the channel.

This was an exciting time in media, at the dawn of the millennium, with the internet a growing force, and citizen journalism gaining power. Zoltan could travel solo with his MiniDV camcorder, filing low-budget yet captivating reports from far-flung places. He was 29, getting paid to travel, to talk to interesting people and to uncover unheard stories—as an agent of one of the world's most recognizable and respected media brands. He wrote web articles and did video for the web, as well as pieces for TV. About half his *Nat Geo* web stories were picked up by the *New York Times* Syndicate, and appeared in publications around the globe, in various languages. Sometimes he was assigned stories, but mostly he pitched his own. The job had freedom, adventure, challenges, and was great for pulling women.

Zoltan's *Nat Geo* work is a tapestry portraying a world on the brink, shown in the tales of unique characters. In La Paz, Bolivia, he reported from the Witches' Market in the world's highest capital. Across the border in Paraguay, Zoltan went searching for traditional medicines in the forest. He tracked down a renowned local shaman, a living library of medicinal knowledge. As-yet-undiscovered cures for diabetes or malaria or cancer could be lurking in the forest. But they would never be found if the tree-felling continued. Paraguay had one of the highest rates of deforestation in the world.

It wasn't only nature facing devastation. Zoltan recorded the poverty among a group of Paraguay's indigenous people, dispossessed of the land they came from. He filmed an indigenous boy getting high breathing into a bag filled with shoe glue, before the boy went to beg by the roadside. Hundreds of indigenous kids beg on busy streets, at the windows of passing cars, looking for money for more glue. Zoltan followed a local journalist fighting to change the plight of the indigenous.

On the other side of the world, he told the story of villagers hunting for leftover American bombs in Vietnam's demilitarized zone: "Ho Hun, a farmer, kisses his infant son good-bye and begins his trek to the Truong Son Mountains near the former DMZ...Carrying a shovel and walking in old flip-flops, he looks like any one of Vietnam's 60 million farmers. But Hun rarely farms anymore.

"Like hundreds of other villagers along the DMZ, in the Quang Tri province, Hun is digging for live American bombs, which he will recover and sell."[2]

The profession of bomb-digging became a thing here after the Vietnam War, more lucrative for many than rice farming. Young men head off to work in the jungles and mountains, shoveling the earth looking for US bombs dropped during the war, which never exploded. They sell the steel casings for decent money. Some hunters can afford cheap metal detectors; others just have a shovel and a keen eye.

Of course, the worry is that a shovel will strike a bomb or a landmine and cause it to explode. But that's the risk these entrepreneurs take. In the video segment for this story, an unusually jumpy Zoltan wonders if a round, clay-caked object might be a mine—Ho Hun whacks it against his shovel to prove it's a rock. "After an hour, we get lucky and find a small bomb, worth about $1," Zoltan says in the voiceover.[3]

At the time he made the piece, more than 5000 people had died in central Vietnam since 1975 from encounters with unexploded ordnance. But finding a good bomb could also improve a family's fortunes.

When Zoltan is asked by media interviewers how he discovered transhumanism, the landmines in Vietnam often figure large in the story he tells. The knowledge and the fear that his life could be blown up at any moment...

Some people run from conflict; some people run toward it. Very few sought passage to Kashmir, the region between Hindu-majority India and Muslim-majority Pakistan, fought over by the two countries since the partition of British India in 1947. But that's exactly what Zoltan did in his *Nat Geo* days. Fourteen journalists had been killed there since 1990. Conflict between armies and militants loyal to the Hindu or the Muslim cause had created unimaginable suffering, with half-a-million refugees driven from their homes, and deaths on a daily basis.

Arriving in Chituk by taxi, Zoltan saw houses turned to rubble. He met a family whose roof collapsed on them while they were eating dinner. They all got out in time, luckily. But other villagers had died in the attack. The locals claimed rockets had been fired by Pakistani forces from the nearby mountains, along the Line of Control. This is how it was now in many parts of Kashmir. Bombings and the fear of bombings drove men, women, and children from their homes, leaving villages deserted. People fled everything they knew rather than sit and wait to be killed.

Zoltan recorded everything he saw. That was why he was there: to bear witness to the plight of the Kashmiris, "pawns" in the struggle between India and Pakistan for control of the land they lived in. These people were caught between the actions of politicians, armies, and militants—all fighting for different destinies for Kashmir.

Nowhere Zoltan went, whether as a reporter or an explorer, cast as long a philosophical shadow over the rest of his life as Kashmir. Amid the torture and death of friends and family—the destruction of their homes and their economy—people yearned to speak of what had happened and begged the outside world for help. Zoltan met refugees who had lost their possessions, livelihoods, and limbs; Kashmiris struggling for normality and happiness. And yet his interviews and experiences suggested that the decades-long conflict was as intractable as ever. Both countries had too much invested in "winning," apparently content to cause any amount of human suffering, and suppress the truth from being told.

While Zoltan was traveling through the region, and when he eventually departed, he hid his camera tapes in his hiking boots. He kept dummy copies in his backpack in case it was searched. He switched hotels night to night, changing up his movements so he was harder to follow.

After journeying across Indian-controlled Kashmir and venturing into Pakistani territory, he was met off the bus in Muzaffarabad by police with assault rifles. They ordered him into the back of their vehicle, and drove to their headquarters.

The fact that he was an American made him a target, but also served as a shield. Not long before, *Wall Street Journal* South Asia Bureau Chief Daniel Pearl had been beheaded by terrorists in Pakistan. No one in authority wanted harm to come to a US citizen and risk the wrath of an American response—and so the police captain gave Zoltan armed guards for the duration of his travels in the Pakistani-administered area. Of course, that also meant they could control what he saw and did. He was followed by a pair of secret police officers for good measure.

Zoltan made a documentary from the footage he gathered, *Pawns of Paradise*, in which the above scenes play out. Though he also wrote stories on his time in Kashmir for *National Geographic, Pawns* is much more focused on the political conflict, which is not *Nat Geo*'s beat. It screened at festivals throughout North America and Europe, and was eventually sold to *The Age* newspaper in Australia. (His footage was later included in an archive collaboration between the US Library of Congress and one of America's largest broadcasters, GHB.) Zoltan's stoic narration of his journey in the region contrasts with searing imagery of families displaced by ethnic cleansing and bodies mangled by war.

After his arrival in Pakistani Kashmir, Zoltan went to a refugee encampment in the mountains near Muzaffarabad. Once well-off farmers and traders were living huddled together in tents, driven from Indian-controlled Kashmir, having been beaten and tortured by the Indian army, they said. Men and women alike were in tears. Some showed deformities from bones broken during torture, or by the explosion of a landmine. One woman testified to him: "When I was at home alone, the Indian army seized and searched my house. They were looking for militants. When the Indian army didn't find any militants, they took me outside and beat me with their guns. My whole body aches now, I have a hard time to even get up."[4]

The tragedy of Kashmir was made worse by the idea of what could be, and what ought to be. Zoltan rowed through the green-hued canals of ancient Srinigar, imagining the glory of Kashmir's summer capital in the time when Marco Polo visited — a glory long lost in the struggle for control of Kashmir's beauty. The Victorian houseboats that glide through the canals attracted as many as 500,000 tourists a year until the 1980s. Now that industry was decimated and locals struggled to survive.

Zoltan trekked on horseback up dirt tracks, past poppy fields, over the hills toward white-clad mountains... Kashmir was rich with natural resources: the world's second-highest mountain, K2; huge forests and raging rivers... People he met described Kashmir as a "paradise on earth" — and argued this was the reason the conflict between India and Pakistan was so fierce and entrenched. Neither side wanted to give up paradise. Whoever won control of the valley could exploit it to bring in millions in tourism dollars. And yet, while the fighting continued, both sides were denied even a share of the prize.

It was a prize apparently worth risking the fate of the whole world for. Every few years, the skirmishes threatened to boil over into outright war between Pakistan and India. Each could potentially use its nuclear weapons against the other. What if a hot war broke out? China—which also controls part of Kashmir—could be sucked into the conflict on the side of its ally, Pakistan, while America came to the aid of India, the world's largest democracy. Soon four nuclear powers are duking it out, millions are dying, with the fate of the entire planet and the human race in the balance. No less than these were the stakes in Kashmir... while the ordinary people there continued to suffer as pawns in the geopolitical game.

Zoltan traveled through shantytown after shantytown; past untended fields with concealed landmines; to a Pakistani hospital near the Line of Control, where farmers missing parts of their legs spoke of the moment they stepped on mines laid by Indian soldiers... All the while, the threat of even greater destruction, through nuclear war, loomed in his imagination. The more suffering occurred, the more positions on each side seemed to ratchet up, rather than calm down in the face of such horror. A compromise for mutual benefit, rather than mutually assured destruction, was impossible to procure. People wanted to die fighting each other for a piece of land—kill countless innocents in the crossfire—rather than change their beliefs. The outside world let it go on and on, other nations reluctant to act because it wasn't their citizens who were dying; perhaps also because Kashmir had no strategic resources, it was just a place of beauty. This was humanity in action. When the stakes couldn't be higher, this was how humans chose to act.

This harrowing interlude in Zoltan's life shaped his own evolving beliefs. He traveled not just across Kashmir but to the capitals of India and Pakistan, Delhi and Islamabad, too. The Indians and Pakistanis he met all wanted, at their core, the same things: peace, prosperity, to see their children thrive and outlive them. And yet, because one country was mostly Hindu and the other mostly Muslim—because they had different views about what happens to the human after death, when in reality there's no evidence anything at all happens, except that your body rots and your mind ceases to exist—this was used as an excuse to wage war and claim parts of the earth belonged to one set of people more than another. It was the most absurd nonsense, with the most awful consequences for millions.

Our species had to be better than this, Zoltan concluded. We must be more concerned about the survival of all. People were dying, losing every chance to build a livelihood and become happy, because governments fought over territory. It was always this way in human history—but it didn't have to be...

***

The memory of Kashmir—and other tragedy-struck places he'd visited around the world—brought home to Zoltan the privileges he enjoyed, and how fragile life's gains could be. Individual humans were so often at the mercy of government decisions, natural disasters, economic collapse.

So, if you see an opportunity to advance and grab a little extra security for yourself and your family—take it. It's not just about you—although there's nothing wrong with wanting your own comfort and happiness. You're in a better position to help others when your circumstances are secure.

In short order over the next few years, Zoltan "grew up"— big time. Even though he loved his journalism work more than anything, he decided a journalist's life wasn't secure—financially or otherwise. It couldn't go on.

Returning to the US full time, he gave up working for *Nat Geo*, and also gave up another gig he had as a communications director and filmmaker for conservationist group WildAid.

Around the same time, he spotted an extraordinary opportunity in the area where he lived, around southern Oregon and northern California. He began putting to use all the DIY skills he'd learned as a solo sailor maintaining his boat, in order to make his fortune.

At the time, the lending environment and the burgeoning property market in Oregon/California were such that he could put $10,000 down, buy a dilapidated house for $100,000, fix it up in three months, and sell it on for $200,000. He didn't have to do anything spectacular for the property to double in value—just make it livable and nice. You can up the value of a home by 10% with a coat of paint and some basic gardening, he says. Turn a large closet into a home office—add $5000 in value. Knock down internal walls to transform an older house into an "open plan" modern dwelling. Every little change adds up to a big payday.

Zoltan figured that if he did this a few times over a few years—keeping costs to a minimum by doing most of the labor himself—pretty soon he would have a million dollars. That was a $50,000-a-year income for 20 years; and more, for longer, if he invested wisely. Plenty for a decent lifestyle.

And so, this is exactly the plan he followed. New Eden Development was the name of his business. He flipped nine properties—mostly homes, some commercial—in south Oregon and northern California, in a span of three years, 2004 to 2007. He picked the most distressed or dilapidated properties he could find, in otherwise good areas: the cheapest with the most potential.

"I hated real estate—you have to understand. I *despised* real estate," he's keen to stress. "It was awful; it was awful work, you know, hard work—and I just didn't like it. After going from *National Geographic* to *that*—it was like there was no meaning." The meaning was the money. His time transforming shitty houses into profitable assets served its purpose. He sold most of his properties at the peak of the market—just before the bubble popped—and made about a million dollars. "I haven't really worked since," he says.

He banked some of the money, invested some in stocks. In the years to come, he continued to buy properties around the US—and ultimately around the world—to keep and sell as investments, and for rental income. But his time solo-flipping was short-lived.

Just three years of "real" work... and Zoltan was set up for life.

The next phase of his plan for a secure and comfortable (but still exciting) existence was... to find a wife. Once he put his mind to it, this happened pretty quickly too.

"I started dating almost exclusively doctors," he says. Medical doctors had the qualities he was looking for: driven, hardworking, ultra-smart, independent, and dependable. By the same token, their field of expertise was different than his, so there wasn't rivalry. Dating became a deliberate, clinical process of elimination.

Within a year, he'd found a match—on *Match.com*.

"Lisa was somebody that, if you meet, you of course marry her, because her qualities are too good to turn down," says Zoltan. "Lisa's math added up on so many different levels."

What first attracted him on the dating site is that Lisa Memmel's profile had no picture. She let the letters after her name speak for herself: BA, MS, MA, MD. That wasn't exactly planned. She had just set up her profile, "and before I had even attached my picture to it, Zolt had emailed me," Lisa says.[5]

Lisa grew up in small-town Wisconsin, the daughter of German immigrants. At 17, she decided to go traveling and figure out what she wanted from life. She spent a semester in Australia, visited Japan, and journeyed through Europe. She entered a Master's program in molecular biology, then went on to medical school—she'd worked out by this point that healthcare, helping patients every day, was what she wanted to be doing. After medical school she worked in Nicaragua and Guatemala, and then Kenya and Ghana. She'd seen a lot of the world, just like Zoltan. Upon completing her residency at the University of Washington and a fellowship at the University of Chicago, she arrived in San Francisco to take up her first permanent job as a doctor.

During a week of telephone calls—before their actual first date—the pair bonded over their mutual love of science and travel. Zoltan admired her ambition, and Lisa loved his crazy mind. They sussed out that they were in the same place in life: ready to commit and settle down.

Lisa eventually took a chance and drove her red convertible Mini Cooper up to Brookings, seven hours from San Francisco. Zoltan's parents' vacation home here was now their permanent residence. But they were away this month and Zoltan was staying there and writing, working on his novel. Lisa's move was a gamble. Her friends had told her not to do it. But she went anyway. They spent a few days together, and Lisa left. Zoltan joined her again a few days later, going to her place at the marina in San Francisco. He brought his cat, Ollie, whom he had adopted from the streets.

Not long after they met, they planned to get married. And Lisa was pregnant before the wedding. The couple married at City Hall in San Francisco on December 28, 2009. Their stainless-steel rings cost $20 each.

Before the big day, the couple had come to a sort of unwritten prenuptial agreement, setting the parameters of their future. "My wife and I had a long talk," he tells me, "and I said to her, 'I'm not working.' Because honestly, I had money." He laughs. But Lisa, who had medical school debt, appreciated the nest egg Zoltan came with.

Zoltan had already started planning and drafting a novel about achieving immortality through technology, a work he thought would become his magnum opus, and he had decided that he was going to devote a big chunk of his life to this idea. He had been contemplating it ever since Columbia, and he now finally had the time and headspace... Lisa had to be OK with that. She was. In part because, at least if he didn't have a job, he would be around to take care of their child while she went out to work.

"He was very honest and upfront when we first met," Lisa says. She laughs. "I have no idea what transhumanism is. And he goes into this long, long spiel. And I'm like, 'He is a little crazy. I think he's crazy in a good way, though. We'll see.'" She adds, still giggling: "The jury's still out on that one."[6]

Lisa had once told herself that she wanted to be married to the most interesting man in the room — and Zoltan fit that idea well.

Lisa didn't have any attachment to or particular interest in transhumanism, even as it became increasingly vital to her husband's life. And he was fine with that, as long as she let him pursue his interest. Lisa's only philosophical red line was that her partner could not oppose or do anything that questioned the provision of abortion services, or any aspect of her work as a doctor. As an OBGYN for Planned Parenthood, she believed to her core in the right to choose. "If you plan to ever get in the way of it, you can plan to be divorced," is how Zoltan sums up the rule he was given. Reproductive rights are "something I'm very passionate about," Lisa confirms. "After seeing a lot of other settings throughout the world, where women don't have any reproductive rights — it really kind of hit a nerve in me."[7] Fortunately, as a libertarian and advocate of bodily autonomy, Zoltan was about as pro-choice as they come.

After Zoltan's many wandering journeys, his reluctance to settle anywhere, his restlessness and waywardness, he'd now committed to a family and to a foreseeable future. His next adventures wouldn't be of the same order as the last. They wouldn't take him on journeys all over the globe — but, finally, on that journey inward to tame the far reaches of his imagination.

Chapter 5

The Bet on Life

"This is what Jethro did today…"

This, Zoltan says, is how he would begin conversations with Lisa about his novel, and the journey he was crafting for its main character, Jethro Knights, throughout the years he spent writing the book. Each evening, Lisa would come home from work at the clinic, and her husband would explain how Jethro's story had progressed. Zoltan's life while he penned his Great Futurist Novel was idyllic. He was able to stay home while his wife went out to work. And Lisa knew that if the novel did well, her husband would be devoting his life to his crazy ideas.

And they were crazy to her. A doctor and a practical woman, Lisa was neutral to skeptical whenever the conversation turned to transhumanism. She listened avidly as her husband described the unfolding plot of his novel… but it was hard to see it as a blueprint for the future.

And while it might have been fun for Zoltan to end every day this way, Lisa didn't welcome it at first. "I remember at the time being like, 'Oh gosh, OK, Zolt's gonna read his stuff again tonight and I have to think about it!'" But after they actually got into a discussion, she says, she found she enjoyed it as a welcome diversion from the stresses of her day. "It was fun and refreshing to hear about where his mind was, the entire day when I was at work—when he was sitting in a coffee shop, madly writing away."[1]

The novel that would see publication as *The Transhumanist Wager*—the work that launched Zoltan's transhumanist career—brewed for a long time before Zoltan actually put fingers to keyboard. A lot of the book was based on his life. For example, the opening scenario was a fictionalized account of a powerful storm he'd lived through in the South Pacific on *The Way*. The boat had been submerged for what felt like an eternity before it righted itself, with Zoltan coughing and spluttering and gasping for breath and life. It was one of those moments—like in Vietnam, like in Kashmir—where death had seemed so close, a tangible enemy that could be defeated if we made ourselves a more durable species. *The Transhumanist Wager* was full of ideas Zoltan had been mulling over, going back to college, his sail trip, even before. It was a lifetime in the making, and the novel would, in certain respects, tell the story of Zoltan's life, with hero Jethro as his avatar, a character who ultimately achieves what Zoltan still hopes to: immortality through science.

Wager is a book with wild hope at its core: Zoltan's dream to live forever. The hope comes from a fear we all face: the specter of death and becoming nothing. Humans have found myriad ways over the millennia to cope psychologically with the fact of our own erasure. Some believe a part of us lives on forever, elsewhere. Some believe we are reincarnated and do it all over again. Some take comfort from the knowledge that family or their society continues after them, that they are but a cog in a machine that will keep on working. Some believe that life is a brief flash of light in an otherwise dark universe, and they're just happy to experience that moment of light.

Zoltan rejected those answers. He chose to believe that we are at a unique moment in history, where the most fundamental changes to the nature of life are taking place, driven by human activity. Not the destruction of the environment and our fellow species on Earth, though that was clearly happening (his time with *Nat Geo* and WildAid had shown him that fact)... This moment also demonstrated that science might soon have a cure for human death.

Zoltan's book is his lengthy answer to a single question. A question he's asking himself—and the reader, too. A question that comes from his studies in philosophy, as well as his life in the world...

What would you be prepared to do to stop your own death?

In *The Transhumanist Wager*, Jethro Knights puts everything on the line in a bet on his own life. *Wager* is heady stuff. As the story begins, the United States—and the world—is in crisis. The economy has tanked. Terrorist attacks are frequent— American Christian extremists target scientists. Chaos and hopelessness reign. Two forces are competing for the future, to set the philosophical direction of the country. Scientists and entrepreneurs hail the potential of genetic engineering, artificial intelligence, robotics, cryonics, and other transhuman fields. They promise new economic opportunities, new possibilities for what our species can become, for living longer—maybe not dying at all. But the transhumanists are up against the status quo. Religious leaders do not want humanity to achieve immortality—it would betray God. Politicians want to regulate all these advances to protect special interests and existing industries, in the name of the common good. More and more jobs and livelihoods are lost, and the stalemate over what's best for America, and the planet, goes on. Ordinary people—caught between these two forces—are scared of big changes that could mean the end of their way of life. But that way of life is already ending.

Jethro Knights is the wild card that no one predicts. In college, he wasn't liked by most students or staff. He's cold and analytical, and at the same time a loner-adventurer. He takes his sailboat round the world, with 500 books aboard. He goes to work in war zones as a roving reporter for *International Geographic*. He likes the mental thrill more than the physical. He acts as if he's the only being who's actually alive, and the rest of the world, whether rock or tree or human, is an inert object for him to study and manipulate to his advantage. If a person or a thing isn't of value to him, he rejects it. Every encounter is purely a transaction that will either enhance or diminish his life. He's obsessed with death—the great enemy that could take away everything he loves, his life and his universe. He seems to be stalked by it in every choice he makes. Facing storms and waves... stepping on a landmine. Death is the end. And there must never be an end.

Jethro's forays into transhumanist philosophy start early. His university senior thesis is called "Rise of the Transhuman Citizen." He lays out a set of beliefs he calls Teleological Egocentric Functionalism (TEF). The philosophy is simple: Seek omnipotence. An individual who lives by TEF, an "omnipotender," must always be growing their abilities and power. To accomplish that, you must live forever. To accomplish that, transhuman science and technology are an imperative— make sure your mind can "live" in a computer, that your body can be regenerated; the method as such doesn't matter. *You must do everything to ensure your own individual survival—eternal survival. To survive in the universe is to control it.* Become God.

The ideas of TEF are distilled into Jethro's "Three Laws of Transhumanism":

1) A transhumanist must safeguard one's own existence above all else. 2) A transhumanist must strive to achieve omnipotence as expediently as possible—so long as one's actions do not conflict with the First Law. 3) A transhumanist must safeguard value in the universe—so long as one's actions do not conflict with the First and Second Laws.[2]

Jethro is mocked and shunned by his classmates. During his defense of his thesis, he's expelled from the room by the professor, who's affronted when Jethro expresses disdain for God as a guiding light. He's affronted by Jethro's lack of empathy and respect. Jethro knows, however, that the end of his university career is not the end for his beliefs.

Jethro's philosophy is not abstract for him. It's a way for him to live his life. If you want to move the boundaries of survival, you have to live on the edge. To know and push the limits of life, you have to go places and do things others wouldn't dare. While circumnavigating the globe in his sloop, *Contender*, Jethro visits lost tribes in Vanuatu, goes volcano boarding in Papua New Guinea—and he writes it all up for *International Geographic*: "Jethro Knights' involvement with the media—perhaps the most powerful social tool on the planet—was underway...His stories inspired the imagination of readers. It made them feel like they were right there, alongside Jethro, sneaking up on armed wildlife poachers or skimming down an erupting volcano on a sandboard," the novel says.[3]

Reporting from the warzone of Kashmir, Jethro meets Zoe Bach, a trauma surgeon on the frontline. They share support for humankind upgrading itself with transhuman science. All ego himself—a man who can't stand not having control; who has little sense of others' worth, outside of how it benefits him—Jethro is transfixed by Zoe's selfless Zen. She accepts whatever the universe has in store. She fights to preserve the lives of others, but doesn't ultimately care whether she herself lives or dies. Everything in the universe is just something in a state of change. Energy becomes matter becomes energy. The notion that someone might be OK with their own death does not compute for Jethro. As they fall in love, they spar.

She tells him:

"I've seen at least a hundred people die in front of me—the ones I couldn't save. And in their eyes, always right at the end—the very end—something happens. Something magical. Something enduring. Something graceful. Like they're going somewhere, or they see something. I don't think it's over, whatever it is that happens."

...

Zoe turned to Jethro and asked, "Aren't you worried you'll miss something if you don't die? Something possibly amazing? You—the explorer who sails the world, and reads everything he can, and wants to leave no stone unturned?"

"I doubt there's anything there, afterward," answered Jethro. "Otherwise, it would hardly be worth it to call myself a transhumanist."

"Dying and being a transhumanist have much more in common than you realize," Zoe answered sharply. "Death is the ultimate arbiter of life, a perfect expression of the soul of the universe. Perhaps death is even the ultimate journey for the transhumanist to undergo. Accepting death and where it leads has nothing to do with not being a transhumanist."

Jethro sighed. "You know Zoe, I don't really understand your issue with death. You seem obsessed with it."

She looked at him, shocked.

"My issue? Are you being funny? Look in the mirror sometime."[4]

When Preston Langmore, president of the World Transhumanist Institute, discovers and reads "Rise of the Transhuman Citizen," he contacts Jethro. The pair begin emailing back and forth. Jethro is drawn deeper into the transhumanist movement, where he calls on those who believe in transhumanism to rise up and achieve the world of their dreams—through violence, if needs be. Governments that make laws restricting scientific advancement; religious authorities that place primitive texts above future technologies; anyone who stands in the way of immortality—they are all fair game. After all, lives are at stake. Death gets closer every day transhumanists refuse to be bold.

Jethro establishes his own organization, Transhuman Citizen, to advance TEF and the idea of the omnipotender. Jethro's breakout moment comes when he catches four Christian terrorists in the act of blowing up a cryonics facility.

He catches them on hidden cameras he's set up. It happens to be the same facility where Zoe works. He broadcasts the footage while talking up transhumanism, and Transhuman Citizen takes off. Young people join in droves. TC organizes strikes and demonstrations wherever transhumanism is challenged. The fightback against religion and regulation has begun.

But it's not enough. The squeeze against transhumanism only gets tighter. The president of the United States creates the NFSA—the National Future Security Agency—a colossal government bureaucracy designed to regulate and influence emerging science and technology, which in practice works to crush transhuman ideas and entrepreneurship. The agency is headed by Gregory Michaelson, a charismatic, popular, well-heeled senator from New York who was once a collegemate of Jethro. The puppetmaster behind it all is Reverend Belinas, America's most revered cleric, who also secretly leads the terrorist campaign against transhumanism.

Jethro has had enough of America. If the US doesn't want transhumanism, he doesn't need the US. He goes in search of an undiscovered country transhumanists can call their own. There must be a way to buy or to build such a place, where research can go on unencumbered, and they aren't burdened by those who oppose them. "Transhumania," he will call this new country...

But before he can get there, a bomb goes off at a transhumanism conference—a blast meant to kill him. It doesn't. It takes out his wife, Zoe Bach, and their unborn child.

"I'll come find you," Jethro whispered when he saw Zoe departing life, unable to control himself, speaking the language she understood.

"Yes...my love...I know you will...I'll be waiting."[5]

***

Zoltan's "Author's Note" at the back of *Wager* states that the book "is the result of two decades of thought and inquiry into transhumanism and the quest for scientific immortality. I wrote it hoping to change people's ideas of what a human being is and what it can become."[6]

In the decade-and-a-half since he was first introduced to cryonics at Columbia, he hadn't spent all his time reading up on it or other futurist topics, by any means. He had done no more research on it, really, until he started to write his novel about conquering death. This was when he first encountered the word "transhumanism." Ever since he'd read there was a movement that wanted to end death, he'd wanted to one day commit his life to it, when he had the résumé to be taken seriously and when he was ready. It was a cause powerful enough that it could consume his life, and that he'd want to convince the world of. He knew he wasn't going to become a scientist or invent an anti-aging drug—there were hundreds of thousands of people who could do that. Zoltan, with his particular interests and skills, would be the prophet, the preacher, the propagandist for a world without death. And not just without death—a future where our capabilities continually increase; where we could live out in space or in a virtual reality; colonize the oceans or become conscious fields of energy. A prophet of infinite power.

Zoltan wrote the first draft of *The Transhumanist Wager* without doing much research into this field with a multi-decade history. He had read only one book about the fields of science that focus on lengthening human lives, Stephen S. Hall's *Merchants of Immortality: Chasing the Dream of Human Life Extension* (2003). Mostly, in writing the first draft of *Wager*, Zoltan focused on what he wanted to be true, and drew on what he knew already. Obviously he was, in a general sense, well read. And he had insight and wisdom from a life of unique endeavors. He knew what he knew from college and travel and years of diverse reading, on history, evolution, Eastern and Western religions;

existentialism and Nietzsche and psychology; great leaders and failed states. He brought all that with him when he sat down to write. He didn't—yet—need to know much about genetic editing or the backstory of cryonics or the nature of artificial intelligence. He didn't need to read up on the transhumanist movement at all. He simply knew these things existed, knew there were people who wanted to end death, knew we could one day have robot body parts and genetically modified brains... and he let his imagination run wild. He didn't like to look into the topics too much before he wrote about them, so he wouldn't be influenced by other authors. He preferred to get his own story, about the future he wanted to see, down on paper first. And then fill in the gaps with research when the need arose. That way, his voice and viewpoint were always center stage. The research to back them up came second—rather than Zoltan acting as a compiler of other people's ideas.

Zoltan hadn't spent his time since college reading about and researching scientific immortality, the species of the future... But he had spent those years thinking often about what those things meant to him; mulling over the philosophical concepts. After his first semester at Columbia, sailing near Borneo, he remarked in his journal: "Most of the desires and propensities of man are not relevant to today—they are relevant to 10,000 years ago. I want to create desires/propensities that are relevant to our age."[7]

He wrote while at Columbia: "Life is a choice of Godhood or dust." Almost exactly the same line is found in *Wager*: "Life is essentially a choice between pursuing personal godhood or dust."[8]

Another note made at Columbia hints ominously at a revolutionary future: "Keys are not the only way to get into doors—there are axes, dynamite, and bulldozers."

The point is, the seeds of what grew into *The Transhumanist Wager* can be found in Zoltan's personal writing two decades before. In his notes for *Wager*, he writes that he built TEF out of his passion for existentialism at Columbia. "I loved existentialism, but I needed something more."[9]

Wager didn't grow on its own from Zoltan's imagination — the book derives a lot, not only from philosophers in history, but also from masters of fiction and page-turner plots. One master in particular.

Jethro Knights was Zoltan Istvan, and also Howard Roark. And John Galt, the hero of Ayn Rand's *Atlas Shrugged*. Jethro shared much of the same background and many details from Zoltan's life. But his character traits borrowed heavily from Roark in Rand's *The Fountainhead*: steeliness; single-minded focus on his own view of the world and his own goals; unwillingness to compromise any part of his vision; a lack of care for other people outside of what they can do for him. These traits are also shared with Galt.

Jethro's relationships with the other characters in *The Transhumanist Wager* also come straight from *The Fountainhead*. In Rand's novel, Roark attends college with Peter Keating, an ambitious social climber who falls under the wing of Ellsworth M. Toohey, an influential socialist writer who helps set America's agenda. Keating has no mind of his own, but he becomes a successful architect by designing boring buildings that others think they like simply because it's the prevailing fashion. Toohey champions Keating's work in his newspaper columns. The reason is simple: Toohey wants our cities, our built environment, to be uninspiring. Nobody should give buildings a second thought. The less inspired people are by what they see every day, the less inspired they'll be to achieve. The less they want to achieve, the more content they'll be with their lot. And when people have accepted their fate, they are compliant and conforming citizens — the kind needed in a socialist society.

Howard Roark's raison d'être is the bold, the inspiring, the never-before-seen — this is what he brings when he designs his buildings. You look at a Roark building and you think of the genius of humans; you think, *Wow — if only I could do something like that...* and you're inspired to follow your own dreams; to achieve the best version of yourself and go as far as you can go in this life. Individuals who are ambitious in this way — selfish; who care about achieving their dreams more than the welfare of the masses — are a threat to Toohey's view of the world. When Roark starts to get commissions, Toohey marks him early as a genius whose reputation must be destroyed. In *The Fountainhead*, the greatest threat to Roark's career comes from Dominique Francon — a newspaper columnist, in league with Toohey, who is in love with Roark. Francon tries to destroy Roark because she thinks a man of integrity like him cannot survive in today's professional environment; she thinks she will be doing him a favor by breaking him and forcing him to accept his fate.

In *The Transhumanist Wager*, Jethro Knights attends college with Gregory Michaelson, an ambitious social climber who falls under the wing of Reverend Belinas, America's foremost cleric and religious conscience, counsel to presidents and celebrities and beloved of the people. Michaelson has no mind of his own, but he becomes a United States senator because of his family's pedigree and because he'll say whatever is popular. Belinas hand-picks Michaelson to head up the NFSA. The reason is simple: Belinas needs a stooge who will do his bidding, use this new government agency to enforce his religious agenda, stamping out the scourge of transhumanism. Cloning, cyborgs, genetic engineering, artificial intelligence — all these things should not exist. They are an affront to God. And people should live their lives in accordance with God's will — they should not have the option to become younger again, or to get a powerful bionic arm, or to make a child from a fragment of skin. When people have accepted their fate, accepted that only Jesus is the way and the truth and the life — and the only path to living forever is through Him — God and Belinas will be happy.

Jethro Knights's raison d'être is the destruction of all of this—with bold, never-before-seen achievements in science and technology. His writings and his activism are all designed to inspire this new Transhuman Age. You read Jethro's work and you imagine how different, how much more, life could be. We could become our own gods; never bow to another authority again. Radical freedom. Such ideas are a self-evident threat to the power and the worldview of Belinas. When Jethro starts to gain public attention, Belinas marks him as someone whose organization and whose ideas must be destroyed. And yet, in *Wager*, Jethro's greatest challenge comes not from Belinas but from Zoe Bach—the doctor and Zen transhumanist who loves him. Zoe tries at every turn to change Jethro: to get him to give up his ego, the notion that he must avoid his own death at absolutely any cost... to get him to accept that the universe might have an order and a meaning that he cannot perceive. She thinks he would be doing himself a favor if—instead of always wanting to control it—he opened himself up to fate.

The character dynamics in *The Transhumanist Wager* really take their cue from *The Fountainhead*. The whole soap opera is there: the strong-willed individualist hero who wants to explode the status quo (Roark/Jethro); the love interest who is the hero's greatest torment, but also his greatest inspiration (Francon/Zoe); the powerbroker who's determined to destroy the hero, to keep his own power (Toohey/Belinas); the social-climber collegemate of the hero, a useful idiot in the power-broker's agenda (Keating/Michaelson).

The plot of *Wager* takes from *Atlas Shrugged*, too. Even though Zoltan wasn't as enamored with *Atlas*, the central plot of that book was clearly useful to him when he sat down to write his. In *Atlas*, John Galt, a genius inventor at the Twentieth Century Motor Company, starts a strike. The strike begins when Galt learns that all the money the company makes is to be pooled, and distributed to workers on the basis of need—as determined by a vote of the entire workforce—rather than on the basis of ability. Galt refuses to accept this regime and walks out, leaving unfinished an electric motor that harnesses static electricity from the air—a form of clean power that could revolutionize the world. Without Galt's genius, the motor can never be finished. Galt's strike does not take place at the Twentieth Century Motor Company, however—but all across America. Galt persuades, one by one, other men and women of exceptional ability to stop working. At least, to stop using the attributes that are uniquely theirs, to stop using their minds—stop inventing new inventions and coming up with new ideas for how things can be done; stop fixing the world's problems. Galt's rationale is that their ideas and their abilities are being stolen. Somebody creates a faster train service, and the government insists that service reduce its speed, so it doesn't threaten the business model of the older services. Somebody figures out a way to get oil out of the ground, to create heat and electricity to keep people warm and literally bring light to their lives; the government nationalizes the operation because it wants control over the distribution of electricity, or to stop a magnate becoming rich, in the interests of the common good. Galt's call to his fellow men and women of ability is: Don't play the looters' game. Let's see how long the rest of society survives when we refuse to do their thinking for them. Galt and the other capitalist heroes gather in a valley hidden by a ray screen in the mountains of Colorado, a valley nicknamed Galt's Gulch. Here, these productive men and women live life as they want it to be: there is no government; there are no taxes; there's no social welfare and no "moochers." These men and women are free to be inventive and to trade freely with each other, value for value, their goods and services. Without these highly productive individuals—the "Atlases" who hold up everything—it's only a matter of years before the outside world implodes. No new inventions; old, creaking technology with no one competent enough to fix it... No wealth being generated to produce the tax revenue to support society... Governments try to cling to power but people fight among themselves for the supplies and scraps that are left... The global economy collapses; a new Dark Age dawns. But Galt and the strikers are ready to return to the world and recreate civilization from scratch.

In *The Transhumanist Wager*, transhumanists cannot be themselves—the government and the religious authorities attempt to crush scientific progress toward immortality at every turn. The only escape for Jethro is the same path taken by John Galt: Drop out of the world, and get others to follow.

***

In the midst of Jethro's transhuman journey, something very human and ordinary—and in its own way extraordinary—took place. Zoltan's first child was born.

"Honestly, taking care of a baby really doesn't affect writing a novel or editing a novel," he says. "Because babies sleep, and... Lisa has a job where she's home actually a lot too. Maybe I was stuck on baby duty four hours a day, but even then I still had 12/13/14 waking hours to dedicate to the novel, so it was a great life. And I enjoyed really hanging out with my kids too."

Zoltan's second daughter was born in the year after *The Transhumanist Wager*'s release.

"I was very lucky, in life, to get four years to dedicate to a single item. A single item where the best of my brain could come out, and the best of my passion... It was such a luxury to take four years of my life and to sit in nice coffee shops and sip coffee and think on how to write something that would ultimately spawn the beginning of, I suppose, a movement... I believed it was going to do that, I believed it was a complete manifesto."

Zoltan had started creating *Wager* before he met Lisa. But it was the stability and routine of married, family life that allowed work on the novel to flourish. He was done with the pursuit of other passions: He could dedicate his time and energy to crafting the novel. He was compelled to.

Even raising his family and making time for his wife were secondary. When I chat to him, he recalls the details of book-drafting far more easily than he recalls the details of his daughters', Eva's and Isla's, first years. And yet, that night we met in Dublin, Zoltan was eager to talk about his family and to show me a video of his kid his wife had just sent him. His dedication to family was self-evident. But he knew his priorities. The highest was to find a way to live forever.

Family finds its way into *Wager*, in a sense. Zoe, Jethro's one true human love, is a doctor—one not afraid to travel to and work in dangerous parts of the world—like Lisa. But the character herself is not Lisa, he stresses.

Despite their more than a year sailing together, and the romantic and spiritual places they went to together, Zoltan's romance with Jennifer Hile was not an inspiration for the Zoe/Jethro dynamic, either.

The woman who inspired Zoe was someone he had a serious relationship with in the couple of years before he met Lisa. Although, it's also true to say that Zoe was influenced by a combination of women and discussions he had with them. When he was on the San Francisco dating scene, he says, he met "so many Eastern-minded women." Zoe's challenge to Jethro came in part from those dates, as well as his own inner critic—his interrogation of his own transhumanist beliefs from his experiences studying Buddhism and other religions and philosophies.

But two years before he met Lisa, there was a relationship that stretched him to his limits, made him question himself and his goals in life. He thought of this woman as he wrote Zoe. The woman's family was incredibly wealthy and close to the levers of political power. This was alluring, but also chaffing for

Zoltan. He says he maybe could have married her if he played his cards right. He thought the world of her. But he knew if he did, behaviors and standards would be expected of him by her family that he might not be too happy to meet. They wouldn't take kindly to the weird futurist stuff he planned to propagate. So, he faced a lifestyle choice: Try to marry into "power" and benefit from its benefits, but lose a little freedom. Or keep his freedom but lose the girl he cared for deeply.

The relationship itself was a constant challenge—to his ego, the way Zoe is to Jethro. The woman was open-minded to whatever Zoltan wanted to talk about, but she also questioned every assumption he made about himself, like Zoe. It was intoxicating and exhausting. Zoltan moved in with her to her apartment in Palo Alto, but weeks later the relationship imploded. He was humbled by the experience, he says, but also set free.

In the end, supportive skepticism in a partner was a better fit for Zoltan, as a person. As an author, he couldn't have Jethro's love interest be neutral on transhumanism; that would never have worked for the plot.

Aside from the central two, the other characters in *Wager* don't have much of a basis in his real life. They are chess pieces in his grand fantasy about how a transhumanist world comes about, more important for what they symbolize in the story than for how we might relate to them as characters. Belinas represents religion as transhumanism's ultimate opponent. Michaelson, the elevation of the mediocre through an appeal to the popular—the temporary eclipse of real genius by a fad, which will ultimately fail so that real genius can rise (the exact same dynamic that occurs with Keating and Roark in *The Fountainhead*). Langmore represents a father figure for Jethro, just as old-timer architect Henry Cameron does for Roark—linking the hero with a lineage of great achievement.

Zoltan admits that "some" characters borrow "a bit" from Ayn Rand, acknowledging Michaelson's debt to Keating, and

Belinas's to Toohey. And he says: "I did seek in some ways to write a modern version of *Atlas Shrugged*."

That the arch-villain of the story is a cleric merits a little further comment. There are Christian transhumanists... Mormon transhumanists... Muslim transhumanists... Buddhist transhumanists... Mormonism, in particular, has a special relationship with transhumanism, thanks to the work of, among others, Lincoln Cannon, co-founder of both the Mormon Transhumanist Association and the Christian Transhumanist Association. Mormons' particular theology, which argues that God was once mortal before attaining immortality, and that we can follow this example to become immortal ourselves, is an easy fit with aspects of transhumanism.

But for Zoltan, atheism was fundamental to his transhumanism. You can, of course, believe there is a god, believe there is an afterlife, and also believe in enabling humans to live as long as possible, and in upgrading their capabilities so they become transhuman. There are people who marry traditional religious views with their transhumanist views. There are plenty who believe that humans advancing themselves does not negate the divine.

Zoltan, however, saw organized religion as an obstacle transhumanism must overcome. Specifically, it must defeat the Judeo-Christian framework that was everywhere in the US. If you believe it's wrong to mess with the biology God created, you believe transhumanism is wrong. If you believe what's in the Bible, and think you should obey God, then you believe it's wrong to seek godly power, so you believe transhumanism is wrong. If you believe you will be rewarded in eternity, then you probably care less about eradicating suffering, and living eternally in the here and now. If you believe faith and the word of Jesus are more important than scientific inquiry, then you're in active opposition to transhumanism. Zoltan wanted to turn all these things on their head. That's why the arch-villain in *Wager* is a reverend, America's foremost cleric—symbolic of an entire culture.

Of course, perhaps the primary question about the characters in *Wager* rests with the main character: To what extent is Zoltan Jethro? To what extent does Zoltan agree with what Jethro does, and think he's a model for a great individual, or for how to go about changing society? When Rand created her hero, John Galt, she was vocal in her view that Galt represented her ideal of what a man should be, possessing all the traits we as human beings should admire and incorporate into our own lives.

Is Jethro Zoltan's ideal transhuman? It's a question that looms large when readers confront Jethro's actions at the end of the novel...

***

On the South Pacific Ocean, three curved, gleaming skyscrapers reach for the heavens.

This is Transhumania, a colossal floating platform—a "seastead"—upon which rests the future. A new nation, its very geography created by man. A city-state, populated with the best humankind has to offer, and growing in power by the day.

Jethro Knights's activist group, Transhuman Citizen, was first established in Palo Alto, Silicon Valley, California. That's where Jethro wrote and published his groundbreaking *TEF Manifesto*, an expansion on his college thesis, where he declared transhumanism to be the next phase of evolution. Today, instead of evolution belonging to the invisible hand of natural selection, humans can take control of our own evolution—this is "our birthright"—and transform ourselves into beings who can live forever and survive beyond the earth.

But the United States of America, once the most innovative nation, doesn't want this future, and neither does any other country. Their leaders and their people are too attached to outdated notions of humanity; to religions that peddle fantasies about an afterlife, and to nonsense doctrines that declare nature sacrosanct—when instead we could use genetics and robotics to become better, and actually become immortal.

Following Zoe's death, Jethro pushes himself even further into his work, turning bitterness at the past and at humanity into action. He decides he must leave this world behind, and instead start a new world. Transhumania will be a new nation established on the open ocean, beyond the limits of existing countries and their laws.

With an investment of ten billion dollars from a Russian oligarch, Frederick Vilimich, Jethro sets about designing and building Transhumania, hiring Rachael Burton, the foremost creator of floating platforms. Frederick, whose wife and son are dead, and Jethro share a mad hope, grounded in far-off science. Founding the transhuman nation, even gaining immortality, are just the groundwork for what might one day come: "If people lived long enough, with all the achievable technological advancements in a thousand years, teleportation into multiple dimensions via antimatter would be possible—and with it, the ability to reverse time and bring back anything anyone desired."[10] Even long-dead loved ones.

The vast ocean platform Burton constructs is home to

three circular skyscrapers of different heights, interconnected by multiple sky bridges. Each building [was] covered in varying hues of glass siding that resembled a computer chip's inner circuitry. During the day, the skyscrapers would mirror their surrounding environment: the sky, the sun, the clouds, the ocean, the city, and the people. During the night, the towers would light up and blend together, forming a brilliant mountain of luminosity.[11]

One of the gigantic towers is dedicated solely to science, another to technology, and the third to housing and recreation:

Everything the nation built and produced would astonish and lead in functional innovation. [Jethro] was determined to create an extraordinary environment like no other place on Earth, where creative human enhancement and life extension research was the highest goal and motive. Where everyone was someone, and the best in the world at what they did. He wanted the best transhuman scientists, technology innovators, computer programmers, medical doctors, and researchers. He wanted the best engineers, designers, builders, artists, and philosophers. He also wanted the best military experts and weapons specialists to defend the nation.[12]

Jethro courts these best and brightest, many of whom share his frustration at being held back; they leave the normal world behind for new lives in Transhumania. Jethro runs his city like a tech startup. He's unquestionably the boss, but citizens are free to innovate and think for themselves, indeed encouraged to do so, as long as they're working toward the goal of the city: a species upgrade.

"People of the world, do not mistake us any longer as citizens of your countries," Jethro declares at a press conference announcing the existence of Transhumania, "or as participants in your societies, or as people who would consider your gods, religions, histories, and cultures as something important. We are not those things. Nor are we willing to accept others' ideas of power and control over us anymore. Nor do we give a damn about your opinions, your social idiosyncrasies, your glam media, your hypocritical laws, your failing economies, or your lives—unless you can offer us something in return to make us give a damn."[13]

Transhumania does offer something back to the old world — or at least, select inhabitants of it:

For those who couldn't afford to get to Transhumania or to purchase its superior medical care, but were worthy to receive it, Jethro Knights created the Immortality Grant. It promised to treat, at no charge, 500 non-Transhumanians from around the world every month, if they were afflicted by life-threatening diseases or debilitating health situations. To qualify and be accepted, an applicant simply had to prove in a short essay why he or she was worthy to receive the free help, but couldn't afford it...

...

...Most of those people were not worthy of the uniqueness of Transhumania's life-saving gifts.

"Here's a Las Vegas lady requesting the grant; she has five kids from three different marriages," announced a young biochemist. He was sitting at the conference table with Jethro Knights and ten other Transhumanians on that month's Immortality Grant team, scouring through endless applications. "She says she's unemployed, lives in a trailer park, is barely able to feed her family, and now has been diagnosed with brain cancer. She says she wants to live longer so she can teach her kids how to be responsible, upright people."

A computer engineer sitting next to Jethro grumbled loudly. "Isn't there a way to screen idiots like that from the applicant pool? What a waste of our time. Send her six feet of rope to hang herself."

"Negative," said Jethro. "The cost of the rope isn't worth it."

"Finally got a good one," a nuclear physicist blurted out. "This 22-year-old Cambodian kid started a small nonprofit group to put solar panels in isolated villages near the border of Laos, where there isn't any electrical power at all for a hundred miles. Unfortunately, both of his legs were blown off by a landmine while on the job last year, and now he can't physically do the work anymore. He's requesting new legs."

"Put him in the finalists pile," said Jethro. "He sounds like the kind of person who would enjoy a few weeks here while we bolt on our newest bionics to get him back to work."[14]

On Transhumania, innovation happens quickly, in healthcare, robotics, AI, organic-tissue regeneration, and weaponry, with the top names in all these fields working independently and together, in an entirely free and supportive environment...

Of course, the eyes of the old world are upon this strange startup, and becoming envious of its growing strength—eager that the upstart be brought to heel.

Eventually, world leaders convene in Europe to sign a declaration demanding that Transhumania submit to monitoring and regulation. They retain the right to use military action to force the new nation to comply.

Jethro Knights does not submit to anyone... The path to war is set. A war that Jethro will ruthlessly win.

***

After all the great writing Zoltan had read, he put a lot of pressure on himself to produce a "great" novel of his own. By that standard, *Wager* falls short—at least in this reader's opinion.

Zoltan makes classic mistakes everybody who takes a fiction-writing course is told not to do. He tells us explicitly about his characters' traits and motivations, rather than showing us through the story. Where sometimes we're crowded with detail that's inessential and distracting—the technicalities of sailing, for example—there are far more important details that are left out. We never learn the first or full name of the novel's major villain, for example. The book is derivative. The character relationships and the plot are basically *The Fountainhead* and *Atlas Shrugged* mashed together.

In the back-and-forth for this biography, Zoltan and I sparred quite a bit about the quality of *Wager*. I should say that my view is not, of course, shared by the novel's many fans, who love everything about it. And I've arrived at my view as a person with a Master's and a PhD in traditional English literature, and someone who likes his own style of writing — with all the biases those things entail. I certainly encourage you to read *Wager* and make up your own mind. Objectively speaking, Zoltan is a much more successful writer than I am.

However, Zoltan recognizes some of the problems I mention, when I put them to him. At university, he was "never a star in creative writing," he says. He was trying to make a modern-day version of *Atlas Shrugged* — the similarities would often be commented on in reviews of the book, and indeed became a major part of his marketing push, as he sought to attract Rand fans to his vision. Beyond that: "I realize the writing isn't as strong as it could be." Of course, Zoltan worked as a professional writer, for *National Geographic* and others, for a long time. But during that phase, he also worked with professional editors, employed by the publications where his articles appeared. "Unfortunately, the book never had that kind of editing." He remarks how professional editors can take his writing — his pieces for *Nat Geo*, the op-eds he writes now — and reshape and sharpen his language, so his thoughts are crystal clear and shining on the page. "I wish I could say I'm as beautiful a writer as Ayn Rand... Maybe I have some better ideas than her, and that is enough to change the world on."

The Transhumanist Wager never achieves the level of depth, clarity, or originality that the scope of its ambition demands. And yet... It's a book that you will never forget. The vision it has for humanity and the transhuman future is revolutionary — as revolutionary as a guerrilla manifesto or a religious text. Revolutionary, in that it calls for a revolution. For people to become believers and take up the cause. And the vision is presented with uncompromising force.

Wager is gripping—especially as the conflict between the transhumanist vanguard and the "coalition of the status quo" comes to a head. The central question is compelling enough that you'll remember, long after, how it played out: How far would you go to stay alive?

Still today, there's very little about the book that Zoltan would change, he says. He's fiercely proud of it, and it reflects his vision.

The book's flaws make a lot of sense in the context of another comment from Zoltan: He wrote it for the first AI. For a machine intelligence thinking about efficiency and optimization, not a human heart seeking to be moved.

Zoltan certainly poured his own heart and soul into the project. He spent a year writing his magnum opus, and three years revising and editing it, full time. The first 15 paragraphs were rewritten 200 times. The published book is a twenty-third draft.

For two decades, the book's idea had germinated without a word being put down. Zoltan began drafting it while he was on the San Francisco dating scene. Every time he broke up with a girlfriend he would go traveling, he says. The first words of what would become *Wager* were typed in Costa Rica— that special country where his mind had first opened, at age 16—after one particular breakup, as he sat in a restaurant with a glass of wine. The words were not the opening of the book, but belonged to a speech where Jethro attacks old-guard transhumanists for their passivity, their lack of urgent direct action toward the accomplishment of immortality; their speechifying and philosophizing but unwillingness to take on and demolish the status quo. He still feels that way about transhumanist organizations today, Zoltan says. "They're like the Diet Coke of transhumanism."

When he started dating Lisa, she was always busy with work—which suited Zoltan. He could do his favorite things: travel and write. On a solo surf trip to New Zealand, he put together an outline of the book and started a draft from the beginning. He recorded his thoughts on audiotape as well as in a doc on his laptop.

Once the stability of married life kicked in, the creation began in earnest, as he added flesh to the skeleton. A lot of the first draft was written in a coffee shop, The Grove, near the apartment he and Lisa shared in San Francisco, before they bought their current home in Mill Valley. He spent hours on end there with a coffee, a laptop, and later, a baby. He kept his earphones in and spent most of his writing time listening to Tool, once described by a critic as "the thinking person's metal band. Cerebral and visceral, soft and heavy...western and eastern...they are a tangle of contradictions."[15] If Tool's music was an attempt to integrate such contradictory forces, so was *The Transhumanist Wager*.

When the novel was advanced enough, somewhere around the thirteenth draft, he printed it for the first time and went through the pages with a red pen—trying to look at them as an objective critic would.

Zoltan was not an expert on transhumanism when he began the book. But as he put meat on the bones of his draft, with extensive research on cryonics, quantum mechanics, seasteading, robotics, and more, he developed an expertise.

The strengths of the book, though—and Zoltan's strengths as a transhumanist writer and speaker today—do not lie in evidence-based science. They lie in the force with which he presents possible futures. In that respect, he's not so different from countless science-fiction writers. The difference is, transhumanism is a philosophy, a way of living in the real world, and of approaching problems. Zoltan's transhumanism exists where science fiction, science, politics, and philosophy intersect.

It offers a glimpse of a utopian future, a view of the future as seen from the present, a possible future if certain current trends continue—in a similar way to *Star Trek*, or Aldous Huxley's *Brave New World*. But unlike science fiction which is just a story, Zoltan wants us to adopt his transhumanism as a method for living our lives, and an ideal that underpins our societies— where we harness politics and science to work toward turning that vision of the future into a future for real.

Zoltan's greatest challenge, as he put together his revolutionary manifesto, was what role he would give to violence. What acts were morally acceptable to commit in the name of eternal life. Creating a fictional version of himself, Jethro, allowed him to speculate, to explore and to pontificate— with plausible deniability. Zoltan could be both Jethro and not Jethro, and Jethro both Zoltan and not Zoltan. Zoltan gave his novel's protagonist much of his own biography; and yet, if the author was ever called upon to defend the violent actions of his protagonist—say he ran for office one day, and he was asked if such actions were justified—he could claim that it was only fiction. A thought-experiment.

Jethro was Zoltan released from human emotions, emotions Zoltan knew he would probably never be released from, and maybe didn't want to be, even if he also wanted to be a machine: compassion, empathy, and love for other lives.

When the four years were up, and he had finally completed the book, Zoltan believed deeply that he had produced something special. A gripping novel. A groundbreaking doctrine. All in one package. He couldn't wait to send it out to agents and publishers.

Zoltan was ready for his book to change the world. But the world had other ideas.

Chapter 6

Fight Like an Environmentalist

The Transhumanist Wager is a case study in how to successfully self-publish.

Zoltan launched the finished novel as a self-published e-book and print-on-demand volume on Amazon, on his fortieth birthday: March 30, 2013. He'd spent most of the previous year trying to find an agent or a publisher. He shelled out $1000 on stamps, mailing out hard-copy book proposals and sample chapters—a strangely old-time way to solicit interest in the novel of the future. He thought a major publisher would jump on it.

No agent or publisher knew what to make of the book. There was some interest in the concepts and the story, but not in the novel as written. Zoltan was refused by every single one.

He gave it to his friends to read, literature fans and entrepreneurs around Mill Valley. No one could believe this was what he'd been doing for the past four years. It sucked. It needed a lot of work.

Zoltan realized then that the novel's champion would have to be its author. He had thought he could woo the publishing establishment; that his vision was so bold, a big company was bound to back it. Instead, he'd now be fighting an online "guerrilla campaign" to get noticed.

He hired an editor on Craigslist to help him smooth out the grammar and syntax and catch typos, and got serious about putting out the best version that he could by himself. He designed a cover that depicted him, Zoltan/Jethro, staring at an X-ray image of his skull. The skull and his face appeared opposite each other against a matte black background; each head had a trail of rainbow light behind it. It was like a Pink Floyd album cover.

A "beta" version of the book was released online in January. Zoltan redid some things—including recrafting the cover, in a mild blue and white—before what he considers the official launch, in March.

Even after all the refusals and the negative feedback, he says, "I really wasn't disheartened very much, because I read the book and thought, 'My god, this is amazing.'"

The more he combed the manuscript, the more convinced he was that he had created this generation's *Atlas Shrugged* (the book that was named in a Library of Congress study as the second most influential in Americans' lives, after the Bible).

There are two lessons already from Zoltan's experience that anybody who wants to self-publish can take to heart: Get the best advice you can... and be prepared to ignore it.

Like any prophet, Zoltan's belief in his own vision was either delusional or prescient. Only time and book sales would tell.

***

There are a couple of things about *Wager*, and the people it attracts, that Zoltan is eager to point out to me. The first is that the book does very well in countries where English is not the first language. Readers there care more about the story and the power of the ideas, and less about the novel as English literature.

Another point he makes is even more telling. On sites like Amazon and Goodreads, *Wager* gets emotive reviews both pro and against. It's a polarizing volume, as Zoltan has become a polarizing figure since its publication. Readers find the novel's vision bold and stimulating. Or they find its concepts appalling and contemptible, and its literary flaws insurmountable.

Who are the novel's fans? Well, that gets to the heart of things. "This is a book that's drawing a lot of 16- to 22-year-old males," he says. Zoltan has captured a market among young white men, akin to other "public intellectuals" on the political right like Jordan Peterson and Ben Shapiro.

Zoltan's appeal is quite different, however. Peterson, Shapiro, and other such right-wing commentators appeal to conservative desires. Peterson tells folks to stand up straight, and defends the achievements of Western civilization, while Shapiro polemicizes against liberals. As a philosophical libertarian and a capitalist, Zoltan is in favor of individuals taking responsibility for their own lives and against a lot of "left wing" ideas like tough regulations on tech companies. These views shine through in *Wager*. However, the novel— and Zoltan's writings since—make clear that transhumanism shouldn't belong to one point on the political spectrum. It is therefore not tied to the Left vs. Right culture wars in the same way that Peterson and Shapiro are. Transhumanism should be a range of ideas everyone embraces.

The goal is to become immortal. How we get there, whether it's through the private sector or government research, the particular policies that lead us there, that's all secondary, in Zoltan's worldview. Peterson became famous defending differences between men and women, and for opposing being compelled to use someone's preferred pronouns. But if you believe there are fixed differences between the sexes, and we should accept and live with this reality... Zoltan's transhumanism is entirely antithetical to that. Transhumanism means myriad new identities, as humans modify themselves with animal and plant DNA, robot genitalia, and all kinds of hormone replacements. As Zoltan posted on Twitter in 2020, transhumanists are so "woke" they want to become a new species. So progressive, they want to progress beyond everything we currently are.

Transhumanists are not conservatives, in any sense. Zoltan would erase all of history and every human tradition if it meant in the future we could be immortal.

Peterson and Shapiro and their ilk are classic conservatives, looking for a return to some older ideas. Transhumanism, by contrast, is a radical concept. Far more revolutionary, in many ways, than anything the political left argues for: Transhumanists don't want a new society, they want a new species.

Zoltan does have common cause with some others on the political right in opposing high taxes, government interventions in the market, and the like. At one point in our interviews, he rants about democratic socialist Congresswoman Alexandria Ocasio-Cortez and how her ideas are bad for America. Likewise, Zoltan's literary hero, Ayn Rand, had common cause with certain figures on the American right as she rose to prominence in the twentieth century. But the radicalism of Rand's ideas— unfettered capitalism, government reduced to only its essential functions (police, military, courts), her principled atheism— meant she could not be described in any way as a conservative. Rand said so explicitly: Those who follow her philosophy, which she called Objectivism, "are not 'conservatives.' We are *radicals for capitalism.*"[1] Istvan's is a radical philosophy of transhumanism.

Rand's ideas overlap in many respects with libertarianism. Indeed, she's a founding figure in modern American libertarianism. Still, Rand disagreed vehemently with libertarians on certain points, and didn't hesitate to condemn them. Most obviously: Many libertarians' ideal is a world without government, whereas Rand saw government as vital for supplying order. Istvan has sought to set himself in the libertarian tradition, but at the same time he swerves sharply away from several libertarian shibboleths.

The foundational idea of libertarianism is the "non-aggression principle." You should not, ever, be the first to use force against another. The use of force is only justified in self-defense. Of course, libertarians view taxes — which are compulsory, imposed by government with the threat of force (you can be locked up for not paying) — as a form of aggression. Therefore, in theory, individuals opposed to compulsory taxation would be justified in committing violent acts against the government in order to liberate themselves from having their money forcibly taken. Zoltan intentionally riffed off this when creating a new kind of non-aggression principle (what might be called an "aggression principle") in writing *The Transhumanist Wager*.

For Zoltan, the fact that we all die is a preventable tragedy, because of the possibilities of transhumanist science. Therefore, anybody who is not actively anti-death, who is not working today toward the goal of indefinite life — or at least open to the concept — is an aggressor against his life. If you actively work to hinder the transhumanist sciences, to prevent the goal of indefinite life — for example, you are a politician, and because of your religious beliefs you block funding for genetic research — you are guilty of involuntary manslaughter. Your efforts have helped prevent immortality, and you have therefore literally caused someone's death. If you stand in the way of indefinite life, you are causing the deaths of innocent people. You have committed aggression against life. Therefore, aggressive acts against you — in order to achieve or sustain the goal of indefinite life — are justified. (Zoltan made his view explicit in a column for *Psychology Today*, where he accused President George W. Bush and the Pope of manslaughter, because their actions blocked stem-cell research and access to condoms.)

In *Wager*, Jethro is prepared to kill people — even those not directly trying to hinder his goals, those who've merely stood by — so that he can become immortal. When the world's politicians turn on Transhumania, sending their militaries to conquer the transhumanist country because it is advancing too far beyond them, too fast, Jethro unleashes the apocalypse. Four drones, named *Trano, Cidro, Kijno,* and *Tabno*—words for the building blocks of life (nitrogen, hydrogen, oxygen, and carbon), in the "universal logical language" (Lojban) Transhumania has adopted—are sent out to lay waste to the old world. People die by the thousand, as global capitals are razed to the ground.

When he has conquered the earth, Jethro becomes its dictator. Libertarian politics mesh with authoritarian methods in this new world order. All previous borders and governments are abolished, and the globe is now Greater Transhumania. Private property is protected, however: Police officers' pay is doubled and their orders are to shoot looters on sight. Jethro brings in Randian and libertarian policies, abolishing Social Security, unemployment benefits, welfare, the entire government safety net. Free-market capitalism will be the economic system in every corner of the world. Jethro does provide free, mandatory education for everyone. In the new age of science and reason, ignorance is a literal crime. Those whose knowledge is insufficient are fined, or imprisoned and forced to do hard labor.

"It gets called a libertarian book all the time," Zoltan says of his novel, "but never by libertarians. It's all the liberals who want to pin the book as libertarian." He laughs. "The book is actually pretty highly authoritarian, and the libertarians get that."

The Transhumanist Wager can pretty easily be interpreted as a justification of any kind of extreme action by anyone in pursuit of what they deem to be a greater good. Hitler thought it was justified to wipe out millions in order to create a genetically pure Europe. Stalin killed millions so that he and the Soviet Communist Party could continue to rule. Jethro is prepared to kill millions if it means he gets to live forever. He even states that he would kill his own wife a thousand times if she were the barrier to his goal. This is the efficient, target-focused morality of a computer that *Wager* challenges us to follow.

Zoltan is dismissive of any comparisons to Hitler or twentieth-century concepts of eugenics, the notion that people with some genetic backgrounds are superior or more deserving of life, which he calls an especially stupid idea in the tech age, because soon we won't even be biological. The struggle in *Wager* is between those with the right ideas and those with the wrong ideas. It's not at all about race. He does also say, however, that Jethro's journey is "like asking ourselves, what if Hitler had the right philosophical basis for some of the things he did?" Hastily adding: "OK, he did not. And I can never make any sense of why you would want to kill Jewish people or anything like that. But let us say that in history there was something like that..." Zoltan is by no means the first person to fantasize about how much better everything would be if he became Global Dictator for a while.

He says: "The book is what it is. It's supposed to be a challenge. It's not necessarily supposed to tell you that the world is a nice place and it's going to end with us all singing 'Kumbaya.' It's actually the story of a dictator, who is willing to force life extension down everyone's throats in order to guarantee that he gets it. And he also wants others to get it, but that's not his first priority. And it's really a question of how far would you go—how far would the reader go—to achieve this, if you wanted to do it?

"Jethro's made his transhumanist wager, but everyone else has to decide how far they would go, what their wager is."

I think it's fair to say that most of us would not go as far as killing countless people so that we alone could live indefinitely. If you value your life... If you love the people in your life... If you love the things in your life... don't you want your life to last? Most of us would answer "yes," but not at any cost to innocent others. And most of us are OK with the idea that we get a turn in this world, and then others get a turn. It can't be our turn forever. But then, most of us also believe that our eternal soul, the souls of the people we've loved, survive forever in another realm. What if that realm doesn't exist? What if this is it?

Would that change your mind about living forever? Would you want indefinite life for your children? Zoltan makes the point that Jethro saves the lives of billions, who will now never have to die, because immortality has been achieved, because of what he's done.

And, in Greater Transhumania, even death may not be forever. Those wiped out by *Trano, Cidro, Kijno,* and *Tabno*... In generations to come, they could be resurrected, given a second chance, with advanced science like quantum archaeology, which could allow us to pinpoint moments in time and recreate them, "3D printing" once-living beings back to life. (Zoltan wrote an article for *Newsweek* about this subject, inspired by the death of his brother-in-law to cancer.) None of this magic is possible unless a visionary like Jethro is prepared to force the earth with him, dragging it through hell to get to a starry future.

It sounds like, to me—for all its looking to the future, and all its cyborg logic—a very old and rightly discredited morality: The ends justify the means.

It reads like, to me, the morality of the omnipotender is, at its core—for all its wrappings in modern technology, AI beings, and futurist philosophy—nothing higher than the instinct of the beast, the justification of empires, and the first law of the criminal: I will survive, I will expand my domain, at all costs to others, because my existence is paramount. Might is right. Strength will out.

But, for Zoltan, death is a form of oppression—the ultimate form of oppression. Humans have the power to stop it, and it must be stopped. Perhaps, by any means necessary.

Those who are oppressed have taken up arms to liberate themselves, committed acts of violence against their oppressors, at every stage of history, and later been thought of as righteous when the regime of the oppressors has ended. We know something about that in Ireland, an island ruled for centuries by an occupying power, by invaders from another island, denying the natives the right to determine their own destiny. Those who've taken up arms against the invaders are celebrated in stories and songs. They were fighting to determine their own destiny.

Death is the colonizer of all life. It steals our children from us too early, in the form of disease. It starves us. Even those who've lived well, who could give us so much more of their wisdom if they had more time, cannot escape it. Those who say we should live with death, accept the reality of it—they are its collaborators. We know what happens to collaborators in Ireland, too. Right now, death is determining everyone's final destiny... Overthrow death, even if one person accomplishes it, and anything is possible.

WWJD—What would Jethro do?

***

The initial trickle of sales for *The Transhumanist Wager* was discouraging. More so than being rejected by agents. Zoltan's goal was popularity, after all—for masses of people to respond to his ideas. He would have to force folks to take notice of him.

Here, he had an advantage. He'd spent years around animal rights campaigners and environmentalists while working for *National Geographic* and WildAid. He had strived to get the message out about the death of the natural world. He knew how to be an activist. This was the quality he wanted to bring to his promotion of transhumanism—and his book about the end of the human world.

Zoltan reflected on the PR success of environmentalism, and wanted to copy that success for transhumanism. Environmentalism is a movement that went from nowhere to everywhere within a few short decades.

In 30 years, environmentalism grew from a few thousand committed activists to a multi-million-person global movement.

It moved from the fringe to the center of global politics and the heart of all our lives. We all now recycle, we demand governments act to halt climate change, we want products that don't aid the destruction of ecosystems. Not long ago, only a few voices in the wilderness called for these things. Of course, our awareness has increased as the existential risk has become more apparent. But high-profile events and the work of dedicated activists and professionals have driven the issues up the agenda in dramatic fashion. From Greenpeace disrupting nuclear bomb tests in the 1970s... To the *Exxon Valdez* disaster of 1989, and distressing images of oil-drenched sea-life beamed around the world... To a generation "awoken" by shows like *Captain Planet* and David Attenborough documentaries... activism and media went hand in glove to bring environmental concerns to the fore, and it's worked. These concerns now permeate all our lives. As Zoltan puts it: "Even Republicans recycle these days."

This ascent of environmentalism, a movement he saw rising firsthand, is a process Zoltan hopes can be repeated for transhumanism. Just as concern about climate change and sustainability have moved to the top of First World conversations, moved to the center of our politics, and become the test by which many businesses are judged... So it might one day be for stopping all human death, or reversing the scourge of aging.

The goals of environmentalists and transhumanists do not have to be in conflict. In fact, they might depend on each other, depending on your point of view: Halting climate change may depend on nanotechnology that alters the atmosphere... Genetically engineer our bodies to photosynthesize — produce energy for ourselves just as plants do, without having to eat — and we'll no longer need any kind of destructive farming. *Vox* hit on a truth about Zoltan's agenda when it reported: "Zoltan's obsessions are weird, but so was Al Gore's fascination with climate change in the '80s."[2] There's nothing as profound as an idea whose time has come... and nothing as ridiculous as an idea before its time.

"I see the transhumanism movement unfolding like the environmental movement, which now has billions of supporters," Zoltan confirms. "The modern environmental movement began because of a Greenpeace boat and a trip to Alaska to stop nuclear testing—or at least that's the legacy we've all been taught. Transhumanism is building its legacies right now."[3]

Zoltan is referring to the 1971 voyage chartered by the Don't Make a Wave Committee, a group opposed to US nuclear testing at Amchitka island in Alaska. Calling their mission "Greenpeace I," the activists sailed toward the test zone to try to halt future explosions. Though forced to turn back because of the weather, their direct action gained worldwide media attention, and the cause of peace and harmony with nature grew and grew in support. That first mission was a template for future activism, and not only against nuclear bombs. Greenpeace put themselves between whalers and whales, occupied oil rigs, disrupted politicians' speeches, and more. One 2020 mission involved dropping boulders into the sea off the English coast to prevent boats fishing in protected areas.

Journalists and "chroniclers" were important to this movement from the start. Some of the very first members of Greenpeace were journalists, and the ability to create spectacle that made news, and therefore got the message out, was vital.

As the risk to the natural world from humans increased in the final decades of the twentieth century, organizations less pacifist than Greenpeace took direct action, too. Zoltan mentions that around the time he sold most of the properties he'd flipped, he dated an Elf. That is, a person who subscribed to the philosophy of, and acted in the name of, the Earth Liberation Front (ELF)—whose supporters/activists were called Elves. The Elves are a decentralized collective of "eco-terrorists" using, in their own words, "economic sabotage and guerrilla warfare to stop the exploitation and destruction of the environment."[4] Killing people isn't their MO. But they might burn down a timber company or an SUV dealership. In *The Transhumanist Wager*, Transhuman Citizen, the activist group, is based on the Earth Liberation Front.

Elves believe their actions are justified because of the stakes: If humans don't stop destroying the planet, we're all doomed. It's literally a fight for survival. Zoltan sees his struggle the same way. "Even if we save the planet, we're all still going to die, unless we create indefinite life."

Chain yourself to a tree to stop it being cut down... Blow up a Humvee to protest its impact on the climate... Throw paint on someone wearing fur... These were the tactics Zoltan wanted to bring to the cause of extending human life indefinitely: "I want to take it to the street... It might be worth fighting for, it might be worth becoming violent. It might be worth strategizing, organizing."

What could that mean in practice? Direct action against organizations and individuals that push an anti-immortality agenda, just as eco-warriors take direct action against those who exploit the environment. Maybe, religious groups that campaigned against genetic science could be targets, or politicians whose regulations limit the advance of technology.

Zoltan summarizes: "If we have a culture that kills people, then that culture needs to be killed."

But first Zoltan needed to get more people to read his book. He rallied himself and thought, if he'd spent four years writing it, he could spend four years promoting it, if that's what it took to reach the audience he knew was out there.

What he did, in a sense, was launch a guerrilla campaign on the internet, inspired by the philosophy, tactics, and actions of the environmentalist movement. Not the actual goals of the movement, but the general principles of their approach. It was these methods, drawing on environmental activism, piloted as he began promoting *Wager*, that he would take with him into his 2016 presidential campaign and beyond.

The first principle was to have a defined philosophy and a clear set of goals, which is a response to what you believe are the stakes at hand—namely life and death, nothing less than the entire future of the world.

When it comes to strategy and tactics, Zoltan had learned from his days chasing down poachers for WildAid that every victory is not just good in itself, but can be used to drive awareness and promote further action. So, when you save an elephant, contact every journalist you think will publish the news, so the story gets out there and the public rally to support you. The same applies when you receive a good review of your novel—spam it across the internet, to every group you can, so social media users can't ignore your work and some are inspired to buy it.

And contained within this was another essential principle. Forget about being polite. Do what you need to do. You are trying to overturn the status quo, and that requires things that will make you and others uncomfortable.

These ideas inspired Zoltan as he went about pushing transhumanism on the public, in the form of his own work.

***

Wager's success—and Zoltan's accompanying success as an activist and opinionist—is a classic story of a snowball effect: his own efforts, combined with the self-perpetuating momentum of a hot topic on the internet.

While he was writing *Wager*, he hadn't given much consideration to online media and social media as avenues to promote the book. He hadn't given much thought to promoting himself at all. He'd assumed it would be someone else's job to promote him. Aside from family life, he'd been largely in his own head for four years, honing his ideas.

He didn't really have a grasp of how to use new technology to spread those ideas. Media and the internet had changed a lot in the years since he'd left *Nat Geo*. It was much easier for anyone to put anything out there online now, thanks to blogs, YouTube, and social, but that also meant there was a lot more content, on every topic, competing for eyeballs. Getting your stuff out there didn't mean you'd get an audience. He'd have to learn this process as he went.

He joined every Facebook group related to transhumanism he could find (over 100 of them, he says), and flooded them with posts about his book. He took the same attitude on Twitter, LinkedIn, and other platforms. His account was suspended several times, since he was behaving like a spambot. He made a special appeal to Ayn Rand fans on Goodreads. He annoyed a lot of people. But others were curious enough to click on the links he posted.

He was ready with a smart pricing strategy. He'd started selling his e-book at $3.99, but that was now reduced. If you were curious enough to click on the Amazon links he posted, fully satisfying that curiosity wouldn't cost you much. You could buy *The Transhumanist Wager* for 99 cents, download, and start devouring instantly. He posted that the price was "likely to change soon" — get the deal while you can.[5]

Sales picked up. There were few books out there with transhumanism in the title. Anyone who followed the movement—and it already had a strong following across the internet—would be at least curious about a novel with the word in the title. Some reviews started to trickle through on techno-utopian blogs, and Zoltan was invited onto a couple of podcasts. Zoltan would do any interview, show up on any podcast or YouTube channel that asked him, no matter the slant of the outlet or its tiny numbers of listeners and viewers.

Then he "got quite lucky," in his words. Prominent transhumanist thinker Giulio Prisco felt the novel merited a serious review on an important futurist website, *KurzweilAI*. (Ray Kurzweil, the site's namesake, was a US National Medal of Technology and Innovation recipient, and Google's AI guru.) Prisco's early review is still a definitive critical insight on *Wager*:

> In *The Artilect War*, [theoretical physicist] Hugo de Garis says that the transhumanist drive to develop technologies to transcend the human condition, in particular more-than-human artificial intelligences, is on an inescapable collision course with traditional morality, religion, and social organization. He believes a massive conflict with billions of deaths is bound to happen someday, perhaps in this century. *The Transhumanist Wager* is probably the first novel to address de Garis' doomsday scenario…I really hope Istvan's fiction will remain fiction, but it seems disturbingly plausible.[6]

Popular tech and science site *i09* (part of web behemoth Gawker Media at the time) republished Prisco's review 12 days later.

A break like this was all it took—*Wager* was away. Ninety-nine-cent copies flew off the virtual shelves. The novel topped sales in the philosophy genre, an achievement that would forever give it the epithet, "No. 1 Bestseller on Amazon." Zoltan felt such a rush to see his book surpass Socrates, Richard Dawkins… Ayn Rand.

More reviews followed, including on the prominent *Huffington Post*, where Ann Reynolds called it a "juicy suspense-fest-page-turner."[7] The reviews and bestseller status gave *Wager* its momentum, and ultimately the legitimacy for Zoltan's transhumanism work and views to cross over into mainstream media.

And that was his next step. Having been rejected by the mainstream, he was ready to infiltrate it.

***

It's worth diverting here to consider how alike Zoltan's success story is to that of his heroine, Rand, only on a compressed timescale. Ayn was also scorned by the mainstream before she became mainstream. Rand's breakthrough blockbuster, *The Fountainhead*, almost didn't get published—except that an editor threatened to resign if it wasn't. Rand's ideas were at first repulsed by groups that would later celebrate them, like Republicans and free-marketeers, primarily because of her proud atheism and because of her extremes, because she saw morality in black and white. Rand yearned to be accepted by mainstream intellectuals, to achieve that validation of her ideas, but she was such a critic of mainstream intellectuals—and her ideas went so against their grain—that this was always a tall order. A similar contradiction is at play with Zoltan. He talks about being an iconoclast, yet clearly considers popularity and the attention of mainstream media, commentators, and intellectuals to be a validation. After her success as a novelist, Rand devoted the rest of her career to promoting her philosophy of individualism and capitalism, Objectivism, through essays and commentary, and giving speeches. Zoltan's career with his particular philosophy of transhumanism follows a similar path.

Rand fled the oppression of Communist Europe for the United States, and devoted her life to creating an intellectual and popular expression of an American ideal: enterprise and self-improvement. Zoltan is only a generation removed from Communist Europe, and he has very much devoted his career to similar American ideals, in a futurist context.

In the six months after *Wager*'s publication, Zoltan continued spamming the internet about his book. Sales led to more reviews, more interviews, more sales... And round it goes... By August 2013, there were 100 reviews on Amazon.

Psychology Today, the popular magazine, invited him to write a regular blog for them on transhumanist topics. They saw that this was something in the zeitgeist, humans' evolving relationship with technology, and that was psychologically interesting. He said yes, of course. "The Transhumanist Philosopher" — "the World's First Mainstream Media Column on Transhumanism"[8] — debuted in September 2013. Then Zoltan made his boldest move yet.

He emailed on spec Arianna Huffington, founder of the mega *Huffington Post* news and views site. He explained who he was and what he'd done: written for *National Geographic*, *New York Times* Syndicate... bestselling author on Amazon.com... And he asked for a column. At the time, *HuffPo* was hosting a number of blogs where authors could post by themselves without going through an editor. It was a golden opportunity.

"I don't think she would have gone for it if she really knew what I was going to do," he says.[9] Zoltan planned to use his mainstream platform to spread off-the-wall theories. But, on paper, his profile ticked the necessary boxes: He'd worked for respected media, and he clearly had an audience.

Now he had a regular column on one of the most-read sites on the internet. The platform the *Huffington Post* provided was a game-changer. It wasn't long before he was writing for *Vice Motherboard* too, and other mainstream outlets wanted to interview him, as an expert on the future.

Almost every day now, there was something new he had written or some article he was featured in that he could promote on social media. The year *Wager* was published, references to transhumanism in mainstream media tripled — and Zoltan played no small part in that.

He had some tricks up his sleeve, as well, to ensure his work got clicks and eyeballs. He put "transhumanism" into the title of as many articles as he could, and used the word as many times as possible throughout an article or post. One of his

HuffPo pieces, for instance, contained the words "transhuman," "transhumanism," and "transhumanist" 27 times in the course of a 672-word article—more than 4% of the piece was composed of these keywords. He was gaming search engine results so that when people typed in "transhumanism," they would find something by him. He even occasionally deliberately used various misspellings, so when anyone put a misspelling of "transhumanism" into a search engine, they were still likely to find him.

His articles were provocative and controversial from the start. His very first piece for *Psychology Today* blamed every American adult for the death of every American child:

> Colorful origami paper cranes appeared on a neighbor's front yard last week, as they often do on lawns across America when a child is dying from a brain tumor. The cranes are supposed to be a heartwarming symbol of eternity, life, and good luck, put up by family and friends to support that child.
>
> Today that child died.
>
> ...
>
> Few people want to address the fact that science and medicine are lagging far behind where they could be if adequate resources were given to them. Even fewer people would agree that they are responsible for that fact. But make no mistake: We are all responsible. We are all responsible for the death of that child. We have not dedicated enough of our time, energy, and resources to the advancement of science and medicine. Furthermore, every time we give a dollar to a religious institution instead of to a scientific institution, every time we endorse a politician who cares more about lobbyists than our fumbling national education system, and every time we support our government's trillion dollar wars instead of a trillion dollar war on cancer, heart disease and diabetes, we support the premature death of innocent people.[10]

It was a call for a better future that could be possible in a changed society, and his aim was to bring these issues to readers who'd maybe never thought about them before.

Zoltan's first article for the *Huffington Post* took a lighter note—and this also helps explain his crossover appeal. He wrote about whether it was time to consider setting up a transhumanist Olympics, where athletes could use any kind of technological or medicinal enhancement—bionics, drugs, whatever. He used the example of a sport he knew well, sail racing, where better boat technology had doubled sailing speeds in the decades he'd been following it. Why shouldn't we be able to enhance our own bodies the way we do our boats?

If he could be fun and thought-provoking, he was also deliberately offensive. It was all part of the same strategy: to get folks commenting and sharing, to raise his own profile and the profile of transhumanist ideas. His next *HuffPo* article was a clarion call for atheists to embrace transhumanism, since it's a natural extension of the rationality and logic that nonbelievers already demonstrate, implying that religious people—since they believe in nonexistent all-powerful beings—lack these qualities. He sums up the ideas of modern thinkers like "Russell, Freud, Nietzsche, Rand, Sartre, Sanger, Hitchens, and Dawkins" as "centered on the fact that a successful civilization didn't need to believe in flying pink elephants or other superstitions to thrive": "If you don't care about or believe in God, and you want the best of the human spirit to raise the world to new heights using science, technology, and reason, then you are a transhumanist."[11]

The "transhumanism is atheism" article kicked off a spirited debate on social media. One commenter[12] with a lengthy post considered it ironic, pointing out that transhumanism often seems to rely on faith for its vision of the future—a belief that immortality will come to pass and take us to the promised utopia—rather than the actual reality of the science. It acts more like a religion than its anti-religion followers like to admit. This is a charge often leveled at transhumanism.

Zoltan's "atheism" article didn't just generate Facebook and Twitter engagement. It truly helped launch Zoltan around the internet and around the world. It had an underlying theme—faith vs. science; whether religion and technology are compatible—that had bothered people for centuries, with plenty still willing to give their two cents today. Sharing, blogging about, or commenting on Zoltan's piece gave them a new chance to do it, in the context of a new movement of the future. Zoltan's language—as is typical of him—was so direct and so bold-faced, he was practically begging people to shout back. Again, all part of the plan. Micah Mattix at *The American Conservative* attacked Zoltan's "perfectly horrifying fantasy" of perfect cyborg bodies, while making fun of his writing style and mixed metaphors, such as how Zoltan had described the transhumanist movement as simultaneously "growing" and "exploding."[13] All this criticism and teasing, of course, elevated Zoltan far above his original station—from humble blogger to subject of the elite's scorn, his ideas now debated in a publication "members of Congress are known to read."[14]

Zoltan's piece was soon deemed worthy of a post on commentator Andrew Sullivan's popular and influential blog, *The Dish*, which also summarized reaction and discussion from others. Half a world away, in Ireland, I read that post (*The Dish* was a favorite of mine) and thought that this "controversial novelist," as the post described him,[15] might be worth investigating for the PhD I was doing on Rand's influence within transhumanism. It was the first time I'd ever heard of Zoltan. It led me to read *The Transhumanist Wager* and to take its vision seriously… as posts like this one must have done with many others.

So, within months of the release of *Wager*, Zoltan was being discussed in some of the internet's most influential publications.

Of course, the likelihood was that no journalist now commenting had actually read the book. It was a long book; people working in insta-media don't have time for such things. You only have time to respond to whatever your competitors are posting. But in commenting on the ideas of this novelist, contained in a novel they'd never read, these writers created new readers for the book, even if they never joined those ranks.

***

Zoltan wrapped up his blockbuster 2013 with an attack on one of America's most beloved institutions: Christmas. "AI Day Will Replace Christmas as the Most Important Holiday in Less Than 25 Years," his *HuffPo* column announced.[16] He declared that the forthcoming birth of artificial general intelligence—when machines become self-aware—will prove far more consequential for human history than the birth of Jesus Christ. The piece became the top article in the tech section of *Huffington Post* for a few days, and was picked up by many other websites. This was a pattern with his pieces—the original *HuffPo* articles would get enough clicks and make their way onto Yahoo!, MSN, and more.

The Transhumanist Wager, and Jethro's journey, anticipated in more ways than one how Zoltan's views would be greeted when they were released upon the planet. Religious and conservative sites and commentators condemned him at every turn, including big names like *Breitbart* and *National Review*, where his predictions and opinions were the subject of fear, anger, and ridicule.

It actually seemed that aspects of *The Transhumanist Wager*'s plot might already be coming true. Transhumanist B.J. Murphy posted on Facebook on December 19, 2013:

In July of this year it became known that the Catholic Church of Madrid, Spain had openly declared war against

Transhumanism...They called for violence against, and the kidnapping of, scientists who are using their resources to help combat aging and transcend biological limitations...

...

...There is no real rational reason for conflict to arise between the Catholic Church and Transhumanists; I do not believe that a scenario similarly told in Zoltan Istvan's The Transhumanist Wager should occur, for we are both on similar missions—the betterment of our people and the conquering of disease and poverty.[17]

Zoltan struck a nerve. His columns distilled transhumanism into emotive issues that touch all of our lives: the possibility of Christmas being replaced... sports... new kinds of sex (with robots)... who's to blame when a child dies... This wasn't some abstract scientific theory. Every reader could viscerally imagine how a transhuman world affected *them*.

The politics of Zoltan's agenda were complicated, and didn't easily match traditional left-wing views (social security, government regulation of business, community rights) or right-wing views (fend for yourself, less control on business, individual freedom). His posts about climate change made him an ally of the anti-environmentalist right. He wrote that we shouldn't be too worried about climate change because, once we become cyborgs, we will not need nature to survive. The worst thing we could do is deindustrialize in an effort to combat a warming planet, when we need technological advancement to continue so that we can achieve immortality. His *HuffPo* article on this theme, "Some Futurists Aren't Worried About Global Warming or Overpopulation," originally published February 18, 2014, was picked up by *News.com.au*, one of Australia's leading news sites, owned by Rupert Murdoch's News Corp. The article, pushing business interests over environmental concerns, then ran in many of Murdoch's papers in Australia.

Zoltan had grown more and more skeptical of the goals of environmentalists and conservationists ever since his days at WildAid. That organization's raison d'être was halting the extinction of endangered species: rhinos, tigers, elephants. But what was the point of working so hard to save these species in the wild, when we could save their DNA, and—in a few years' time—clone them en masse, and repopulate the earth with these animals if we wanted?

At the same time, Zoltan's columns spitballed left-wing ideas, like tax increases and massively cutting back on military spending. He proposed a 1% additional tax on everyone in order to fund life extension, and diverting the military budget to science and technology to improve Americans' wellbeing. Even diverting money from prisons to education.

In one sense, his ideas were broadly libertarian; he would describe this as his underlying political ideology: favoring business freedom over government intervention, individual rights over community welfare. But this was really only in theory. In practice his ideas came from all over the political map, and were focused on one goal: whatever it took to achieve indefinite life as expediently as possible. Though he was in theory a free-market guy, he was in favor of massive government spending on science and tech, massive intervention in the market to support his favored industries. Though he was in theory an individual-rights supporter, he was happy to entertain authoritarian ideas like controlling who could become a parent, and monitoring new immigrants with drones, when he thought such ideas might lead to a better society.

A year after *The Transhumanist Wager* came out, Zoltan was still doing interviews about the book. In 2014, *Wager* won some indie-author accolades. It was named a National Indie Excellence Awards (NIEA) runner-up, and a winner in the "Fiction: Visionary" category of the International Book Awards, further bolstering his reputation. His articles were winning big attention. He'd gotten some truly hardcore fans, like software engineer Chris T. Armstrong—a frequent re-sharer of Zoltan's ideas on Facebook, whose mind was blown by *Wager*. Armstrong was now writing a book elaborating on *Wager*'s ideas.

"Zoltan's novel gave me the same positive feeling that I have always gotten from *Star Trek*. Humanity made it! And we didn't just survive, we were finally able to thrive, to excel, to surpass our lowly ape-origins," Armstrong told me in an email.[18]

For every fan like Chris, there were also those within the transhumanist community who were skeptical or critical of Zoltan. The Institute for Ethics and Emerging Technologies (IEET), one of the foremost transhumanist platforms, published a range of articles both praising and attacking the ideas in *Wager*. In a particularly pointed commentary, Rick Searle wrote that, "Istvan, by creating a work that manages to disparage and threaten nearly every human community on earth, has likely *shortened the length of your life*." Since Zoltan is intent on attacking every human tradition in his book, where then does he expect the support for transhumanism to come from? A transhumanism that must turn to fascism as its method of implementation "does not deserve to survive," Searle wrote.[19]

Zoltan's novel and his columns, and the increased profile they gave to transhumanism, initially appeared to be broadly welcomed within the community. But as time went on and the columns became more controversial, the backlash was strong. The main critique being: If this guy is what the world associates with transhumanism, we're fucked. Some, like Searle, believed Zoltan would turn the world away from life extension, because of his extreme views and pronouncements.

Though the transhumanist community itself developed something of a love/hate relationship with Zoltan, he was now increasingly seen in the media as a leader of this fringe movement and perhaps its chief spokesperson. He appeared on Morgan Spurlock's CNN show along with transhumanist royalty, Max More and his wife, Natasha Vita-More. He was flown across America to appear on John Stossel's show, airing on Fox Business and Fox News. (He tells me another Fox News star, Tucker Carlson, even asked for a copy of *Wager*.) And Zoltan had his champions among prominent life-extension advocates, as well as his detractors. Celebrity anti-aging scientist Aubrey de Grey promoted *Wager* on Facebook, calling it "rather splendid."[20]

The Ayn Rand connection helped him cross over from nerdy subculture to more conventional libertarian and business platforms. "I'm thrilled to know that there's been someone that has written a book that's been compared to *Atlas Shrugged*, because I admire all of the concepts that it talks about," gushed Heather Wagenhals as she introduced Zoltan on her radio show, *Unlock Your Wealth Today*. Wagenhals, a personal finance guru, said she reads *Atlas* every year to "sharpen the saw."[21] Their shared Rand fandom opened the door to explain transhumanism to a whole new audience and discuss the financial systems of the future. How does Social Security work when folks typically live to age 150?

Zoltan's biggest story yet was published in *Vice Motherboard* on August 4, 2014. It discussed ectogenesis—growing fetuses in artificial wombs—"a social and political minefield," he acknowledged: "It has the possibility to change one of the most fundamental acts that most humans experience: the way people go about having children. It also has the possibility to change the way we view the female body and the field of reproductive rights."

He went a little into the history and predicted this method would be widely used within 30 years. The benefits were obvious: Women would no longer have to experience pregnancy and childbirth and all its complications for their own health, lives, and livelihoods. The process could also be safer for the baby, which would not be at risk of catching a virus the mother had, for example. Outside the womb, every single moment of a fetus's development could be medically monitored:

> The ectogenesis technology itself is highly complicated, though somewhat simple looking. Basically, it appears as an amniotic fluid-filled aquarium with a bunch of feeding tubes and monitoring cables attached to a live, developing organism. Those tubes bring the nutrients, oxygen, etc needed to grow an organism and help it survive; the cables monitor everything going on inside the tank. There's certainly a *Matrix* feel to it all.

Experiments with goat and mice embryos had already shown promising results. Religious scholars and some feminists were united in their view that artificial wombs could harm women's position in society, as they'd lose a unique aspect of identity. But for those who could not bear their own children, whether infertile women, or transwomen, or male couples, the technology offered new promises of dreams coming true. Zoltan concluded that ectogenesis is "one of [the] hottest topics of the transhumanist future."[22]

And he was proved right. His article went viral. Within days, it had produced 1200 comments in a discussion on Reddit. The article went around the world, with Chinese and Arabic and other translations. The topic was picked up by *Newsweek*, who quoted Zoltan, as did Fox News. Meghan McCain discussed the subject live on Pivot TV. Undoubtedly, the original article excited some committed misogynists, who wondered, with glee, how long it would be until women were no longer needed in society at all. Taking a different tack, countless Christian sites repeated the content of the article with notes of horror and dread—*Won't somebody please think of the children?* What kind of child would it be, who had no natural origin?

Zoltan brought his own experience into the article, mentioning his wife was an OBGYN. His own life experiences certainly lent his writing and his interviews gravitas and credibility—he wasn't somebody with no life experience speculating on things that would only affect others. He'd worked in war zones, he'd built a business, he was raising a family—his views were weird but he wasn't a weirdo. He was as American as apple pie.

He followed up the ectogenesis piece with an even more contentious op-ed on parenthood, this one for *Wired UK*. Should we give everyone birth-control implants, and make them get a license before they can become parents? he wondered. The article aired out in the open a subject more people have probably entertained, at least in their own heads, than would care to admit it. Is it morally right that we can all just become parents by default, no matter how unfit we may be to raise a child? *Won't somebody please think of the children?* The article began with an anecdote that rings true. At a party, "One tall blonde woman said something that caught my attention: with 10,000 kids dying everyday around the world from starvation, you'd think we'd put birth control in the water." Zoltan doesn't say in the article that the woman was Lisa—his wife was his inspiration. The piece goes on: "In an attempt to solve this problem and give hundreds of millions of future kids a better life, I cautiously endorse the idea of licensing parents, a process that would be little different than getting a driver's licence."[23]

He was quickly called a "moral idiot,"[24] and worse, by the religious lobby. It was pretty easy to chuck accusations at him that he was a eugenicist, but Zoltan saw his motives as humanitarian: Children deserve a good life, and the government should vet parents in advance to make sure they can provide it. Of course, practically and ethically, there were so many problems with this suggestion. Who decides on the standards for a fit parent? How can you force folks to be sterile? But Zoltan saw it as a thought-experiment worth doing, given the number of children worldwide suffering from illness, starvation, and slavery each day. Birth-control technologies were the key to implementing a licensing regime like this. For Zoltan, the radical possibilities of technology to improve human life, and make our bodies more than what they are now, included topics like this.

This article is especially noteworthy among Zoltan's repertoire because of how it exemplifies his political thought-experiments. His idea was squarely totalitarian; there was no doubt about that. But it had libertarian roots, in that the responsibility for having and raising children rested firmly with each individual or couple. Part of the goal was to avoid the need for an ever-expanding welfare state and redistribution of tax money. The traditional libertarian position would be that it's no one else's business whether you have children, or how many children you have, but it's never the state's responsibility to support them either. So, if parents can't afford to feed their kids, and private charity can't pick up the slack—as is the case in communities all over the world—then children simply starve; and, morally, that's not a failing on anyone's part, aside from the parents', according to libertarianism. "I don't know if you've seen anybody starving to death—I have," Zoltan tells me, frankly. He thought he could square the circle: Find a way to prevent the suffering of millions of children while keeping parental responsibility as the primary issue; avoid placing obligations to care for other people's kids onto taxpayers. This was the modus operandi of many thought-experiments he was now running: totalitarian policies for the common good with individual rights at their root, made possible through advanced technology.

The article is noteworthy for one more reason, and that is—it brought the attention of the commentator who would become, Zoltan says, his great intellectual nemesis, whom he also calls his greatest promoter: Wesley J. Smith at *National Review*, an attorney and opinionist whose conservative Christian screeds against this and later pieces amplified Zoltan's message exponentially.

The media feed off of each other. So, every mainstream outlet that commented on Zoltan led to more. All of a sudden, "this guy who is questionable, and his questionable book, is now everywhere in major media," Zoltan says of himself. The conservative media "felt it important to counter this transhumanist ideology," which meant he was being taken seriously.

Zoltan had started pissing off everyone. *PJ Media* called his idea "Birth Panels," playing off the "Death Panels" comments used by former Republican vice-presidential candidate Sarah Palin to describe government-controlled healthcare. That article implicitly referred to him as a "wannabe eugenicist on the left" — ironic given where his politics were coming from.[25] Zoltan was now receiving death threats in online comments, his ideas compared to the schemes of Lucifer. (One Christian website later added Zoltan to its list of possible antichrists, citing the fact that his first, middle, and last names all contained six letters: 666. Also on the list were Barack Obama and Canadian prime minister Justin Trudeau.) *InfoWars*, conspiracy theorist Alex Jones's site, featured fear-mongering commentary on his birth-control views — a sure sign that Zoltan had made it into the hated elite.

He laughs now, thinking back on how he used major media to relaunch the transhumanist movement in his image; how he went from cuddly stories about drug-enhanced sports to such topics as "teaching children about Jesus is child abuse." (In a September 2014 *HuffPo* piece, he wrote: "Like some other atheists and transhumanists, I join in calling for regulation that restricts religious indoctrination of children."[26]) He started writing for *Slate*, then *Gizmodo*. He was appearing in print interviews, podcasts, documentaries, at conferences, all the time. He was invited to give a TEDx talk.

The buzz he was generating, the nerves he was touching, fed Zoltan's ego and made him think he could become part of an even larger conversation.

What would Jethro do? For his next step, Zoltan decided to do something that Jethro would never do. Jethro's modus vivendi was to seize what he believed was his by right—he never sought popular support. Jethro would surely look on the idea of running for election with disdain. (Indeed, after Jethro remakes the world as Greater Transhumania, through violence and conquest—becoming dictator over the globe—he decides not to run in the first election he holds, once the order he's established is secure.)

Zoltan was not Jethro. He saw a route to his transhumanist goals through democracy, through building a movement and a groundswell of people's support. He was ready to throw his hat in the ring, to become a contestant in the greatest popularity contest on Earth.

One morning in September 2014, Lisa found a sticky note on the fridge with Zoltan's handwriting.

"I'm running for President."

Chapter 7

The Science Candidate

Robots vs. Dinosaurs—Optimus Prime vs. T-Rex. That's how it was billed: the clash of the transhuman and the primitive. Zoltan vs. Zerzan.

Not long after he officially declared his run for the presidency online, Zoltan took part in a great debate. Well, the event itself was not so fantastic. But what it represented—and the fact that he was part of it—was an important step forward. A battle for the soul of America, with Zoltan on the side of progress.

A student society at Stanford University in California, the Stanford Transhumanist Association, hosted a clash of ideas, a debate between transhumanist Zoltan and John Zerzan, a one-time associate of the Unabomber and arguably the world's leading anarcho-primitivist thinker. Zerzan believes that humans should give up all our advanced technology, even farming and permanent settlements like cities and towns, and return to nomadic hunter-gatherer lifestyles.

The idea for the debate came from a book. Journalist Jamie Bartlett ended his 2014 study of internet culture, *The Dark Net*, by interviewing Zoltan and Zerzan, two figures representing the two extremes when it comes to modern technology: Embrace it wholeheartedly vs. give it up entirely. Perhaps wisdom or a happy medium could be found in analyzing these two philosophical extremes. After they were both interviewed by Bartlett, Zerzan sent Zoltan an email asking for a debate in a public forum: "Are you up for a public discussion or just another coward who can't back up the techno-worship you advocate?"[1] The Stanford Transhumanist Association was happy to provide the forum.

The night before the event, Zoltan posted a photo on Facebook, showing him studying up on Zerzan. He was reading a 2011 interview in *The Atlantic*, published just after Apple founder Steve Jobs's death, where Zerzan argued that in 500 years—if humans live that long—inventors like Jobs will not be celebrated.[2] Industrialization requires exploitation, people working in poor conditions for poor wages: the laborers in China who make iPhones. And those of us who use iPhones become, ironically, disconnected; we lose genuine human connections, with our faces stuck staring at screens. When we need the Baby Cry app to tell us what our own child's cries mean, we're lost. We're forgetting the skills needed to survive on this planet as human beings. By contrast, in a world where all we need are stone tools, almost everyone can do the work— there's true equality. But once we have powerful technology, we need experts to run it, and that gives the experts power over us. This kind of hierarchy is wrong and harmful. This is what Zerzan believes.

For Zerzan, tech is bad because it takes us away from what's human. For Zoltan, it's great for precisely the same reason. Zerzan wants us to give up our technology and return to our original humanity; Zoltan wants us to embrace technology and move beyond our humanity. You could hardly find two people who agree on less. That was the point of getting them to debate—a clash of two worldviews completely opposed to each other. Online, on social media, a promotional poster featured Optimus Prime of the Transformers battling a tyrannosaur—a fun way of representing Zoltan the futurist clashing with Zerzan the prehistoric.

As it turned out, the event was a bit of a damp squib.

Zoltan, in his write-up afterward for the *Huffington Post*, said that "some transhumanists" were worried about their safety, because there were anarchist supporters of Zerzan, dressed all in black, in the back.[3] At the start of the debate, someone kept laughing while Zoltan was trying to make a point—and Zoltan threatened to throw this person out, even though that wasn't really up to him (he was just a speaker, not the event organizer). Zoltan losing his cool and snapping at an audience member was not a good look for a presidential candidate. Not a good sign for his chances to withstand the pressures of the campaign trail.

Since there was literally no common ground between the participants, it was less of a debate and more a case of two people taking turns to describe what they thought was good in the world, what was good about life. There was no moderator. The participants just took turns making statements and responding to each other. Each saw the opposite things as good. Zoltan talked about his visits to primitive tribes—how the Mareki had such a high rate of early deaths; where would we be without modern medicine? Zerzan argued that primitive lifestyles did not exploit the environment, threatening all life. Zoltan argued for all the amazing things humans can do with technology—it's cool to become an astronaut and explore everything out there in the universe. Zerzan said it's not good to become an astronaut; all the industrialization required for that is destroying the Earth. Zoltan argued we need to use technology to free ourselves from human limitations. Zerzan claimed that by wanting our smartphones, tablets, laptops—all our technology—we're making slaves of others, the people who build them for us.

Though Zoltan and Zerzan are extreme examples of arguments on either side, the essence of their debate is ancient: the conflict between the new and what already exists, between progress and the world as-it-is—whether preserving the world as-it-is means the natural world or human traditions, or both. There are always, as Bartlett puts it in his book, those who greet new things with optimism and those who greet the same developments with pessimism:

In Plato's *Phaedrus*, Socrates worried that the recent invention of writing would have a deleterious effect on the memories of young Greeks who, he predicted, would become 'the hearers of many things and will have learned nothing'. When books began to roll off Johannes Gutenberg's press, many suspected they would be 'confusing and harmful', overwhelming young people with information. Although Marconi believed his radio was helping humanity win 'the struggle with space and time', as his invention became popular, others feared that children's impressionable minds would be polluted by dangerous ideas and families rendered obsolete as they sat around listening to entertainment programmes. We don't know if early *Homo sapiens* argued whether fire burns or warms, but you can hazard a guess that they did.[4]

The Stanford event was essentially a debate between the optimism and the pessimism that greets all progress. In that way, it had relevance beyond its own specifics—speaking to a deep conflict within the human species, and one particularly relevant to the Transhuman Age. Zoltan set himself at the vanguard of the new, out to confront and persuade all the forces of the Old World.

The core of the debate really goes to what the human being *is*. What's the meaning of human life? What's the purpose of a human being? What's the "best" way to live as a sentient creature? Should we make it our goal simply to survive, for as long as nature allows, in harmony with others and the environment around us—to subsist? To find food; take care of each other and our fellow creatures as best we can; live and die by the rhythm of a natural order.

Or... Is there more to what it means to be human? Zoltan's progenitor, Ayn Rand, argued—radically—that humans can survive only by changing the natural order. That's how it always has been. We break rocks to make stone tools. We chop down trees to build homes, create shelter. We domesticate horses, dogs, farm animals to support our developing needs. And with all this, we grow. Each generation learns from the last. We are the only animal for whom this is possible, for whom progress is possible. And so, the cave becomes the hovel becomes the skyscraper. The balloon becomes the airplane becomes the starship. To be human is to make progress. To be able to do more, over time. It's what we are.

Rand saw these two views of what it means to be human in stark evidence in 1969, when she wrote about two events that took place within a month of each other: the Woodstock music festival, and the launch of *Apollo 11*. Rand associated these events with two terms used by the philosopher Nietzsche, the Dionysian and the Apollonian. These two terms describe two aspects of humanity: the Dionysian, from Dionysus, the Greek god of wine and fertility; and the Apollonian, from Apollo, god of truth and precision. Those indulging the Dionysian side of life could find a way to primitive transcendence— drunkenness, orgies, excess, the forgetting of mortal things, and finding another plane of existence. The Apollonian, in contrast, reflects how thoroughness and logic can result in outstanding achievement and beauty: a finely made sculpture, for example, accomplished through diligent work.

Rand saw these aspects in the quintessential event of the hippie movement, and the historic launch that took men to the moon. Dionysian Woodstock, revelry without a cause; masses of people seeking a communal escape, to return to a kind of tribal existence, a kind of primitive harmony with all things— something absent in the modern world with its destructive technology. And *Apollo 11*, in contrast, the exemplary achievement of scientific progress, of hard graft in math and experiments over time, folks building on one another's knowledge. And now, humans traveling to the stars, walking on a part of the universe other than Earth—opening up new avenues in the imagining of what's possible...

A raft of other writers have described a similar moral divide within humanity. They see different boundaries and variables, but a similar essential split. For example, the moral binary in *Zen and the Art of Motorcycle Maintenance*, one of Zoltan's favorite books. Author Robert Pirsig tries to resolve the tension between romantic thinkers, who want to be transported by beauty, and in some way see technology as a betrayal; and classical thinkers, who like details and problem solving, and want to understand and indeed advance technology. Zerzan/Zoltan... Dionysian/Apollonian... Romantic/Rational... They are not the same divide, exactly, but they speak to the same question about what's human. What do we value more: the fact that we are animals, one part of a broad universe; or that we are logical individuals, capable of controlling the universe around us—chiseling a boulder into a sculpture; understanding physics to build a motorcycle, or a rocket ship; using technology to extend our capabilities?

The divide goes to the role of the human mind; how we should use our ability to think. We have the capacity to learn and to build on what's gone before—but does that mean we should? Is the creation of electricity worth it, when it puts candle-makers out of business? We face the same question today with countless jobs and AI. Or, put another way: Why would we spend millions to go to the moon or Mars, just because we can, instead of feeding our fellow humans here on Earth?

For Rand and Zoltan, we should make technological advancements because we can. This is our unique capacity as human beings: innovation; altering the natural world; learning to do more things, more quickly (from the assembly line to page-load time)—traveling faster, building higher, living longer, going out to the stars. There is an opposing set of philosophies about humanity, however, that believes the universe, the planet, and ultimately we ourselves would be better off if we retreated from this tendency to control and disrupt, and agreed to a simpler way of life.

Yaron Brook, chairman of the board of the Ayn Rand Institute, one of the primary spokespeople for Rand's legacy, said on his podcast in 2020 that he wasn't a transhumanist, because transhumanism isn't "an ideology," it's just where the future is headed: "In 10,000 years, human beings won't be exactly human beings." We'll have "technology embedded in our own genes" and evolution will be "managed." Humans and robots won't be separate entities from each other. So, why declare yourself a transhumanist? It isn't any special set of beliefs, it's just where things are headed anyway—if tech progress continues.[5]

It's true, transhumanism is not a discrete ideology, in the sense that there are many ways to believe in and support progress. There are many ways to describe that progress and the technological future, from the singularity to the AI Age to simply "the future." But "transhumanism" is nevertheless a useful term because the two parts of the word suggest a particular method and way of working, a direction to orient ourselves. Yes, innovation and tech development are inherent in humanity itself. But we want to use that fact to take it to the next level. We want to become "beyond human."

Besides, Ayn Rand knew better than most that progress was not inevitable. It requires work. The "right" ideas need to be spread—in education, in the media, in politics—and to be believed. A philosophy of pursuing the highest heights needs to take hold across the culture. That's the work Zoltan saw himself engaged in, under the banner of transhumanism. The work of making progress possible. Confronting a culture that says we should accept our fate. Educating folks during on-campus debates. Appearing everywhere he could in the media, spreading ideas. And joining the world of politics, to influence policy. A good explainer of ideas, a good writer, good on camera, good one-on-one, he was the renaissance man of transhumanism: charismatic; humble in person, for the most part; ruthless in his self-promotion. He became a one-man transhumanism juggernaut. That was his strength—and his problem.

For Zoltan, confronting and dismantling certain assumptions in US and global culture was key to overcoming opposition to transhumanism. Not just our "deathist culture," as he called it—the tendency of everyone, particularly religious folk, to consider death "normal," part of the Lord's plan or simply inevitable. Even good, since it meant your loved one's soul was now in a "better place." Zoltan also wanted to confront head-on those parts of the culture that greet technological progress with skepticism or outright hostility. It was a viewpoint he'd seen grow, not only as technology had advanced but also as a consequence of environmentalism and growing alarm about climate change. People were worried about what industrialization, capitalism, and all our tech advancements had done to the planet, and there was a growing tendency to look on deindustrialization or stalling progress as a solution. Anarcho-primitivism offered what was obviously an extreme solution, but it represented a broader tendency within our culture, in its purest form. Zoltan thought it important to debate and defeat the underlying principles behind any anti-technology sentiment, if transhumanism was to triumph.

***

But Zoltan didn't just need to convince those on the opposite side of the technology debate… He also couldn't quite convince many other transhumanists that his pursuit of the presidency was the right way to go.

There's a saying in Irish politics, which comes from writer Brendan Behan: The first item on the agenda is the split. With any new movement, it seems the first thing that happens is that disagreements bubble up about the finer points of ideology, how exactly to pursue the group's aims, who's in control, whatever. And breakaway organizations or rival factions are set up.

So it was with the United States Transhumanist Party. Zoltan announced on social media, on October 7, 2014, that he'd founded this new political party, and considered himself its potential candidate for president.

"There's this idea that anyone can run for president," he says. And it's easy to get in the race—you just go online and sign up to run. Actually having a chance of *winning*... Well, that's a whole other sport, of course. But Zoltan knew he couldn't play that game. Declaring his candidacy for president was about something else. "I was looking for a venue, or looking for a means to essentially reach the most amount of people. And I thought, how do you do that?" Hundreds and hundreds of nobodies run for president in the US every four years, to service their ego or bring attention to a pet cause. Why shouldn't Zoltan do the same?

In his case, however, his ambition went wider: He wanted to grow a movement, and he founded the Transhumanist Party as a vehicle for the movement and his candidacy. The party was "an attempt to generate massive support for life extension science and to get people to embrace general transhumanist ideas," he wrote in a Facebook comment.

The complaints mounted up quickly. Like anyone prominent on social media—and by this point he had thousands of followers—he received lots of unsolicited advice.

"I would prefer the term Transhumanist Counsel instead of 'party', it captures more," wrote one commenter.

"How is Transhumanism, and the right to self-ownership and self-modification, not already covered by Libertarianism?" asked another.

The objections ranged from questioning his process to fundamentally ideological issues:

We don't have a transhumanist in Congress yet, so planning for presidential candidacy without enough constituents nor a firm amount of allies in Congress makes it a moot point.

Politics is religion. Forget becoming grand wizard and change the world organically.

This is one of the worst things possible that could occur for the transhumanist vision...

If you are going to be the President, you are going to handle some war issues...I think [it's] better if no transhumanists are involved in war...

I would suggest running under one of the major party names for Governor... I would also suggest promoting 'Preventative Healthcare Research' instead of Trans humanism.[6]

Along with plenty of praise and encouragement, these were the initial comments he received.

"It was so filled with controversy," Zoltan says of the party's creation and his presidential ambitions. "Everyone attacked me."

Zoltan was not the first transhumanist to run for office. Also in 2014, Gabriel Rothblatt, son of the founders of Terasem (a Florida-based spiritual community devoted to life extension), ran unsuccessfully for the US Congress in his home state, as a Democrat. Italy elected a transhumanist, Giuseppe Vatinno, as a member of parliament in 2012.

If the initial reaction to transhumanism joining the political sphere was mixed, it wasn't dismissed outright. The emergence of the US Transhumanist Party received increasing media attention... and not just at home. An article in the UK *Telegraph* (by Jamie Bartlett) opined that transhumanism was likely to "become part of the political furniture," given how technology was advancing.[7]

On January 1, 2015, Zoltan announced that folks could now become members of the Transhumanist Party. Pretty soon, inspired by the US, Transhumanist parties were being founded all around the world—loosely affiliated, varying in their specific politics, but all motivated by a common core idea, much like the world's Green parties. Zoltan helped set up an online portal to assist folks who wanted to found global parties.

There were plenty of positives happening for the movement, as a result of Zoltan's establishing the US Transhumanist Party. But, as his popularity grew—particularly with the launch of his campaign bus tour in 2015—he also made enemies. The biggest fight with the transhumanist old guard was still to come.

***

How *do* you run for president?

This is something Zoltan spent most of the first year of his campaign figuring out. And he wasn't dealing with the infrastructure of an existing party—he was trying to build that from scratch.

He looked into getting his party officially recognized by the Federal Election Commission, even sitting down with FEC staff in Washington, DC. But he would have needed a lot more infrastructure than it was possible to build quickly. The Transhumanist Party would have needed separate state parties and candidates running for other elected offices. Zoltan would've loved to have Transhumanist Party candidates for statewide office and for Congress—but he didn't have time to go searching for and vetting those people, and helping them run for election. He had to stay focused on the goal of creating buzz in the presidential contest.

At least at first, "running for president" mostly involved Zoltan doing what he was doing anyway—writing articles, appearing at conferences—except now he could say he was the transhumanist presidential candidate, which added an extra level of urgency or curiosity to whatever he was doing.

He did have an official platform, published on the party's website. This included designating aging as a disease and encouraging secularism, to combat religious opposition to transhumanist aims. Reduce the number of government employees through new technology. Legalize all drugs, and divert money from the "war on drugs" into rehabilitation. Pour cash into space exploration. The platform became another bone of contention... of course. The folks becoming members of the Transhumanist Party wanted a say in how it was run and what it would stand for... and this wasn't part of Zoltan's plan.

Zoltan, in his words, allowed folks to join the party as members "for a few days and then canceled it all. It was just a total, total disaster." People not only wanted a say on the platform... They wanted to change the logo... They wanted to change the website... They wanted all sorts of changes. This wasn't the kind of democratic process Zoltan had in mind when he decided to join the democratic process.

He decided to stop accepting members and to explicitly run the party as his personal "company" (he owned the trademark "Transhumanist Party"), a publicity outfit for his causes and ambitions. He treated the party "as a startup," he says, "knowing that many startups fail"—unless they have a strong leader and a core vision pushed relentlessly by that founding figure. He compares himself and the Transhumanist Party to Steve Jobs and Apple. He wanted to move fast and break things. He'd seen other transhumanist organizations struggle to get attention, and didn't want to be like them. He mentions Humanity+, formerly the World Transhumanist Association, in particular. "I watched Humanity+ go from barely relevant to irrelevant."

"This is why startups need a very strong CEO, who just basically acts as a dictator. Because if you have too many people with their opinions, nothing ever gets done. And in the beginning, this is a race to survive, this is a race to make this relevant."

Though the Transhumanist Party and activities surrounding Zoltan's presidential run were largely a solo affair, organized and driven by him, he did have helpers.

Zoltan shut down membership of the party. But he had appointed several voluntary officers and advisors, specialists in particular areas, to promote the notion that this was a party of serious ideas. Aubrey de Grey, well-known gerontologist, became his "anti-aging advisor." Gabriel Rothblatt, the transhumanist who'd run for Congress in Florida as a Democrat, became his "political advisor."

Adding to this, there were several dozen activists on email lists, committed to volunteering at least five hours a week to help the party—handing out bumper stickers, whatever it was. Many devoted far more time. Transhumanist Party meetups took place in several US cities, from San Francisco to DC, though just a few people ever attended.

"Thankfully, I didn't need other people's money to do what I was doing. I had my own money," Zoltan says.

He continued to write attention-grabbing columns, especially for *Vice*, which had become a go-to edgy publication for millennials. When you campaign for president as a political nobody—running for a party that can't get official recognition—there's not much more to be done than write manifestos, and hope they hit a nerve.

"Let the robots take our jobs," he wrote in one of his *Vice* articles: We should embrace automation as a "New American Dream."[8] The robots can do all the hard work—and we can live easier lives. To keep everyone happy, he proposed a universal basic income, supplied by the US government. The Age of Robots meant there wouldn't be enough "jobs" for everyone, but people wouldn't have to hold down a job because they would automatically receive a basic income, and then they would be free to pursue their hobbies or creative projects or other opportunities for income...

Being in favor of automation and a universal basic income—both controversial issues in mainstream politics—were the least controversial of Zoltan's ideas. In one *Psychology Today* column, he argued that transhumanism would supersede and render unnecessary the LGBT+ rights movement. That's because: When we have a transhumanist culture, everyone will choose their own sex organs, their own gender; we'll try out life in all different kinds of bodies. We'll have sex with robots and cyborgs, and no one will fit into minority "boxes" like L, G, B, T, or anything else. His hypothesis, of course, ignores the question of essence: Being gay, or transgender, isn't a choice, it's an identity. Zoltan simply didn't think sexuality or body type would matter in the transhumanist age. He wanted those who were already outside the mainstream—the LGBT+ community—to join him as allies. But, for some of his readers, he was erasing who they were.

Not content with going after the LGBT+ community, Zoltan next took aim at people with disabilities. The headline on another of his *Vice* pieces blared: "In the Transhumanist Age, We Should Be Repairing Disabilities, Not Sidewalks."

The article was a statement of a transhumanist given: that everyone should want to be able to do more with their bodies, their limbs, their minds, their faculties. Upgrade themselves. But he was widely accused in comments and commentary of ableism, denying the basic rights and dignity of disabled people. He was giving a total free pass to society, saying it shouldn't have to adapt itself to the needs of people with disabilities—having bathrooms for wheelchair users, buildings where there was an alternative to flights of stairs, traffic lights that beeped for the blind. Instead of universal accessibility, in Zoltan's view, we should aspire to universal ability. Those with mobility impairment could get bionic limbs; the blind given bionic eyes...

"He presumes to have the right to make the sweeping declaration that all people who use wheelchairs should just be 'repaired,' instead of repairing the sources of the problems that we face," wrote disability rights activist Emily Ladau in a response.[9] One commenter on Facebook put it bluntly: "This is how we are, how we were born, came to be and there are many who are fine with this. It is ableist thinking that our lives are less than yours…and that we should somehow be grateful that there is a [Transhumanist] party that is here to fix it."[10] The Center for Disability Rights published a response by Leah Smith: "Just because a body is disabled, it does not mean it needs to be fixed," she wrote. Let's "fix sidewalks and then let the disabled people decide for themselves if they want to be fixed next or not. This way, they'll be able to, at least, arrive to the appointment."[11]

"Ableism" was a new one for Zoltan. What he hadn't realized is that for many folks with a "disability," it's a part of their identity and culture, like any other element of their lives. They just want equal access to the same things as anyone else—why should they have to change who they are? For example, deaf people have their own language and community; like any distinct group, they value it and want to keep it.

Zoltan was not chastened but emboldened by the attack. He didn't want to eliminate anyone's identity, of course. But like a good libertarian, he didn't think everyone else should have to commit their resources to accommodate others' differences, either. And when he looked at videos of deaf babies smiling upon hearing their mother's voice for the first time, after receiving cochlear implants, he knew he was right to believe in upgrading individuals, not buildings.

Then there were Zoltan's articles that seemed to be about wild sci-fi stuff or a far-off dystopia, but at the same time struck close to home. In "What If One Country Achieves the Singularity First?" he wondered about a near future in which a single powerful nation possesses a superintelligent AI, which could quickly calculate how to improve the welfare of its people while also upgrading weapons capability to wipe out rivals... What if China commanded this awesome power before the US, and the US was all of a sudden a second-world country, at risk of irrelevance or oblivion?[12] For Zoltan, as had been the theme of *Wager*, a simple evolutionary force would always drive one person, or group of people, or a nation, to be top dog, to conquer all rivals, be the ultimate power in the universe. The trick was to be that person or nation. Somebody was going to be. In the end, it was the only way to ensure you weren't conquered. To ensure your own survival.

Zoltan saw the race to create artificial general intelligence as a new arms race, a new Cold War struggle that could secure America's dominance or end it. The government and individuals had to commit resources to this battle.

He also wondered, and wrote about, whether—when artificial intelligence arrives—religions might try to convert AI... An AI could be a new prophet, since its advanced calculations could tell us more and more about the nature of the universe, and predict future events... Would priests try to save the soul of a sentient robot? "I don't see Christ's redemption limited to human beings," a Florida Presbyterian pastor told him in an interview for the piece.[13]

Would you like to win $1 million just for living to age 123? This is another sci-fi-show-type concept that Zoltan promoted. A Moldovan multimillionaire businessman, Dmitry Kaminskiy, pledged to pay this amount to whichever human reached the milestone first and broke the record of Frenchwoman Jeanne Calment, who died in 1997 aged 122, the oldest person ever to have lived. Kaminskiy was trying to get people to think about how they could live healthier, smarter, and thus longer. Zoltan wrote about it for the UK's *Daily Mail*. It was the kind of intriguing concept that could motivate anyone—and that was the idea. It got you wondering what you could do to increase your chances of scoring that million dollars...

What could you do to help yourself live longer right now? That was the practical question that lay behind all of Zoltan's writing—and the question he wanted to provoke. Every reader had to make their own transhumanist wager.

The columns were fine—many went viral. But Zoltan wanted a way for his campaign to reach every American. He felt they needed to hear it. He was looking for an idea that would make him stand out from every other "also ran," make him a viable and serious candidate for president.

That's when he decided to convert a secondhand bus so it looked like a coffin, and drive it around America.

Chapter 8

Roving Metal Coffin

It all started to go wrong pretty quickly. In fact, within the first couple of hours of a planned four-month trip.

They were headed toward the Mojave Desert from San Francisco when Zoltan noticed the bus's engine smoking. Fuck. He stopped the vehicle—and there was a trail of oil leaking from the engine back into the distance.

What the hell was he going to do about this?

There were journalists on board from the *Telegraph* and *Vox*. There was a documentary crew aboard. There were Transhumanist Party supporters riding with him. More media were following behind in another car. Zoltan had promised them all a road trip into the future. He wasn't even sure he could make it a few more miles down the highway. The engine was overheating, and there was oil spilling everywhere. Things weren't looking good for the self-described Science Candidate.

The whole campaign-bus tour might have to be abandoned— after months of planning and a crowdfunding campaign that had brought in $20,000.

He scratched the back of his neck, and looked into the sky. The sun was already setting.

Zoltan had survived on Advil for three days straight while he labored to get the 1978 RV he'd bought on Craigslist—the "Immortality Bus"—ready for the road. He'd transformed this rickety bus into something resembling a giant coffin in his own front yard, much to the concern of his neighbors in the well-to-do San Francisco suburb. He'd hardly slept in the last 72 hours. Now, the coffin-shaped shuttle might be the death of his dreams.

What on earth had made him do this? And how could he get out of their immediate predicament?

Surely there was a way to salvage the situation... But he had to think fast...

***

Bingo! Zoltan thought.

Rachel Edler, a pro graphic designer and transhumanism supporter who was creating the look of his campaign bus, had just shown him a first illustration of a bus outfitted like a coffin. This was much better than the original idea—a bus with a huge microscope on the roof, symbolizing science.

Months deep into his run for president of the United States, Zoltan needed a jolt. The big-time race was now taking shape. Hillary Clinton was on her way to the Democratic nomination, while Donald Trump was about to turn the Republican Party inside out, making it his own. The historic unpopularity of both eventual nominees made third parties—like the Greens and the Libertarians—more appealing to many voters. In this landscape, an independent candidate might grab more attention than usual... That was the hope.

Zoltan needed an idea that would create excitement around his campaign, that would galvanize the Transhumanist Party troops, and suck in more media attention. He also needed a campaign bus.

Why couldn't the bus be the thing that created the excitement? he wondered. It would have to be a special bus. Every politician running for something went around in a bus emblazoned with their campaign slogans and photos of themselves. Zoltan wanted to do something different. What if the bus represented the ideas driving his campaign, in a deeper way? But at the same time, make it zany and fun. To gain any traction at all, third-party or independent candidates in US politics need unique angles. His idea could be something the media jumped on, if it was crazy enough.

Road trips have a revered history in America. And Zoltan was going to take his bus not just on standard campaign stops, but on a road trip to remember. What if he brought the journalists and campaigners aboard to all sorts of bizarre places — biohacker conventions, cryonics facilities, transhumanist spiritual centers? Fringe ideas can break into the mainstream if you make them into wacky adventures that editors and producers judge will bring eyeballs to their content.

Zoltan was inspired by Ken Kesey, whose hippie road trip with his "Merry Band of Pranksters" in the psychedelically painted bus, *Furthur*, became one of the iconic set-pieces of the 1960s, furthering the counterculture. "To some extent we modeled the Immortality Bus off that idea. The idea that a single road trip could transform the landscape of a culture," he tells me. Transhumanism was ready for its breakout moment as a countercultural force, Zoltan believed, and a series of weird adventures on the road could just tip the scales. Whatever grabbed headlines and gained followers.

For the vehicle itself, Zoltan imagined something like the art cars or "mutant vehicles" he'd seen when he attended the infamous Burning Man festival in Nevada's Black Rock Desert — cars transformed to look like animals, or everyday objects, or cartoon characters, or other crazy things. Anything but a boring old car. They were like something out of *Mad Max*.

His first notion was that he could attach a giant fake microscope to the top of his campaign bus. It would certainly stand out. He was the candidate pushing science and technology as a way to remake human relations and American politics, and he wanted to represent that. But the reality was, a microscope bus would cost a lot of money — and the shaft would snap off at the first underpass.

A bus was already roughly the same "box" shape as a coffin. There was another way to represent the ideas behind his campaign: Remind people of the finality of death. Death was the enemy his campaign would destroy. Do you want to end up in a coffin? Or do you want to join Zoltan in the fight against disease, aging, and the grave itself?

It was a suggestion from Maria Konovalenko, an anti-aging scientist and the Transhumanist Party's "life-extension advisor," that first sparked the coffin-bus idea. Konovalenko told Zoltan she wanted to float 400 cardboard coffins down the Hudson River to highlight the transhumanist fight against death and aging. That's when the lightbulb turned on. They should make up the campaign bus like a coffin. It was simple, it was more economical, it was perfect. The coffin was a powerful symbol of where we would all be without science and life-extending technologies; without transhumanism.

That was how it began. Zoltan looked on Craigslist for old buses for sale. He launched a crowdfunding campaign on Indiegogo to help him cover the costs.

Zoltan also spoke to his dad to get his advice, since his parents used to own an RV. The candidate ended up purchasing the same make and model Steven and Ilona had once owned.

Nature had to be defeated, literally, before the bus was ready for the road. The 38-foot-long, 1978 Blue Bird Wanderlodge he eventually bought — for $10,500 — ran, but it needed a lot of work. It also came with three wasps' nests in the air-conditioning system.

That wasn't the only problem. Zoltan had never driven a bus before, and he would have to learn how. Presidential candidates don't usually drive their own campaign buses. But, well-off though he was, Zoltan couldn't shell out for a driver for months, and he couldn't rely on any volunteers to be able to drive the bus either. What about insurance issues? This was his show; he was responsible for taking it on the road. He'd once taught himself how to sail—he could do this.

The next step—now that he had the "Immortality Bus," as he decided to call it—was to make sure there was actually enough interest in the road trip he was planning, to make it worthwhile. He began to map out a rough route for his modern-day *Furthur*. There was a biohacking festival in the Mojave Desert on September 6, 2015, so that seemed like a logical place to start. A big Vegas technology fair was happening just a few days later, and the organizers agreed to let him speak at that. Zoltan and Co. would spend the first leg of the journey doing events around California, Nevada, and Arizona. They'd then head through the South—bringing the message of immortality-through-technology to the Bible Belt—and finally north to the nation's capital. Zoltan planned a three- to four-month trip. That was long enough to draw the right amount of attention to be a game changer. He wanted it to culminate in Washington, DC. He would pen a Transhumanist Bill of Rights—a twenty-first-century Bill of Rights for a renewed America—and deliver it to Congress. The journey was a publicity gimmick, an adventure on the road, and an activist mission, rolled into one.

The bus would have some passengers who were donors to the crowdfunding campaign. And then there was the core group of Transhumanist Party activists whom Zoltan could count on to follow him anywhere. But the people the candidate really wanted aboard his bus were the media. Without media attention, it was going nowhere. He contacted a load of journalists, and got his first couple of bites. Then it was game on.

As he made his plans, Zoltan worried whether the journalists would actually show up. He also worried what his wife thought.

A year before, he'd announced to Lisa that he was running for president by leaving a note on the refrigerator. They'd laughed about it, but then it got serious. "I don't think she realized that there was going to be cameras in our house" on a regular basis, he says.

Zoltan is unapologetic: "We all come to a relationship with our parameters set up. And mine was, I want to be able to pursue indefinite lifespan."

The Immortality Bus was just the newest iteration of the crazy ride they'd been on since their wedding, when Zoltan was already writing *The Transhumanist Wager*. Lisa wasn't a cheerleader for his ideas, but she didn't veto it either, when Zoltan said he wanted to take a few months to travel around the country in a bus dressed up as a coffin—even though they had two young kids at home, and all the pressures of that. She knew her husband was who he was. And she supported him following his dreams, like he supported her career.

It's that live-and-let-live attitude that Zoltan has valued most through his transhumanist pursuits. Because of Lisa, "I've had the freedom to become who I wanted to become," he says.

"Even if it's not always the easiest thing to do... If you love somebody and you want to make things work out, you've got to support them," Lisa says. "You never want to look back and regret that you weren't supportive enough."[1]

Besides, there could be fun stuff for the family too on a road trip. A visit to Vegas?

And a young family was catnip to journalists covering presidential candidates. He would have to get Lisa to join him with the kids for part of the journey. It could be good for everyone.

It was now mid-summer 2015. He had the bus. He had a plan. He had people on board with the plan. Now he just had to turn the bus into a coffin.

***

The fear that it wouldn't get done in time—or at all—was real. The crowdfunding campaign was launched on July 8, 2015. On the Indiegogo page, Zoltan put up a one-minute video to show what he hoped the bus would look like, using sketches drawn by Rachel. He told the viewers: "I'm going to be driving a 40-foot coffin bus across America, promoting transhumanism and life extension. It's something that's gonna work. It's something that's going to wake up America..."[2] With a pledge of $100, you could book a tour on the Immortality Bus at one of its pre-arranged stops (seven people selected this reward level). To ride with Zoltan on the bus, you had to commit $1000 (four big-spenders signed up for this). For $50, you got a signed copy of *The Transhumanist Wager*.

Zoltan was wary about spending more and more of his own money on the campaign. He had the resources to pay for a lot himself, and he was fine doing that, since it was his cause, and he also didn't want to be beholden to donors. The more people contributed, the more of a say they would have—and Zoltan always wanted the final say. But nor could he spend so much that it threatened his family life. If he could have other people pay for the Immortality Bus tour, that would be great. He could keep his resources for the core political campaign; and besides, the Immortality Bus tour wasn't just about party politics, it was an experience to share with people. You weren't just giving money to a political cause; you could join a fun and historic road trip. So, it was the sort of thing more likely to garner donations than a doomed Transhumanist Party campaign manifesto. A successful fundraising campaign also added legitimacy—it was good PR. It showed there was support behind what he was doing.

The Indiegogo campaign sought $20,000. Zoltan achieved that—but only thanks to Dmitry Kaminskiy, the Moldovan multimillionaire and life-extension fanatic mentioned in the last chapter, who previously pledged to give $1 million to the first person to live to age 123. Kaminskiy gave $15,000 to the bus campaign. Most transhumanists don't have much money, Zoltan insists—they're 20-something gamer nerds. As well as that, plenty of longtime activists couldn't get on board with the direction Zoltan was taking. They certainly weren't going to commit their money to it.

The academics of transhumanism and many longtime activists—folks who'd been involved in "transhumanism" far longer than Zoltan—hated the idea of the Immortality Bus, and hated his steering of the party that bore the name of *their* movement. Zoltan paraphrases their concerns: "You are going to take what is a nice academic movement and turn it into a big circus." He admits: "That's exactly what I was trying to do. And the reason is that the circus is going to be 100 times more popular than the academic side of the movement."

It was hard at first to find time away from his other campaign responsibilities—writing articles, giving talks—to focus on what needed to be done with the bus. But it was now summer, and he needed to make decisions.

He looked into hiring someone to make the bus look like a coffin for him. Too expensive. And would they even get it right? Did anyone have any experience dressing up buses as coffins? Better to do it himself, working to Rachel's plans.

The design plan was straightforward enough: Attach wooden slats to the top of the bus on the outside, slanted to look like a coffin lid. Then paint the whole structure, the whole bus, a walnut-wood brown. The pièce de résistance was a giant wreath of flowers perched atop the "coffin lid." Then no one could mistake this for an ordinary old motorhome painted brown.

Zoltan told Lisa that he was bringing the bus to their home, and he was going to be working on it in their front yard. He was under deadline to get it all done in time for the media who'd agreed to come on the maiden voyage (Labor Day 2015), and as usual, the person Zoltan trusted most to do what he needed done was himself.

He'd once kept a boat afloat for long stretches as a solo sailor, meaning constant maintenance… He wasn't shy about physical work. He'd flipped more than a dozen houses... Now he was flipping a bus. He made many trips to Home Depot to get all the supplies.

The first problem was the wasps. They were in the air-conditioning system at the top of the bus. Right where he needed to be working to turn the vehicle into a coffin. "It was very hard to work there without being stung," he says. But he sprayed out the wasps, and plowed onward.

The next problem was the neighbors. Maneuvering the bus through his affluent suburb to his Mill Valley home, "I took out someone's wall," he says. "Crashed into it a bit." Driving the 40-foot, rickety machine was "pretty complex," he found. "I could barely fit the bus in the front yard." He was now laboring all hours of the day, hammering and making a racket as he worked on this rusted, derelict-looking RV in this neighborhood of multimillion-dollar homes. Busily working under the leafy catalpa trees that shaded his front yard, and for which his street was named. "I really didn't want to let people know what I was doing," he says. "I live in such a nice neighborhood." And so instead he was just feeling this weird social pressure to do something amazing with his DIY project.

He changed the tires on the bus; it looked like they hadn't been replaced since the 1970s... He cleaned out the interior... Put on the coffin lid... And finally, there was the paint job. Not everything was finished by launch day. He still had to paint some campaign info on the outside and do some other finishing touches. He hadn't stuck the fake flowers onto the roof. But time had run out. Journalists were arriving from the *Telegraph* in the UK, *Vox*, and Polish media for the maiden voyage— two documentary-makers, too. Two days before launch, Roen Horn—founder of the Eternal Life Fan Club on Facebook, and a big Zoltan supporter—arrived and helped paint the bus. Roen had also agreed to professionally film the entire voyage, despite never having held a video camera before, so Zoltan would have his own footage. Rachel was on hand to do the "Immortality Bus" lettering on the side. "I was blown away by how much [the bus] looked like the concept art," Rachel told me.[3] Zoltan worked three days straight—popping pills frequently—to get something serviceable for the road.

Another problem was the growing sense of responsibility that Zoltan was feeling. Roen didn't have health insurance. He lived with his parents and didn't have a job. What if he—or anyone—got sick on the trip? What if they had an accident? It wasn't at all unlikely in this 37-year-old jalopy. Zoltan was worried the RV couldn't drive very far, and he'd be an embarrassment in front of international media—who'd be sure to write about and report gleefully on the Quixotic misadventure of the transhumanist Icarus, who thought he could transform an ancient bus into a vessel of the future, and with it convert others to his cause. The disastrous headlines were easy to imagine.

Zoltan had captained a boat before, but he hadn't been any sort of leader in any real fashion. He'd tried to keep anyone who was with him on the sloop—and himself—safe in emergencies, but *The Way* was a voyage of pleasure. He'd worked freelance—on his own—for a long time. He'd even built houses by himself. But now he was the captain on a different kind of voyage. He had a crew—his Transhumanist Party colleagues—who were depending on him. And he had the eyes of the world, in the form of those reporters, constantly upon him. He'd never been responsible for other people in this way before. He was glad he was a parent and had that experience.

Labor Day. The day the journalists arrived at his house, Zoltan was still painting. He hadn't driven the bus since he'd taken it on the two-hour journey to his home from where the seller kept it. The coffin top was complete, but he didn't know if it was going to blow off the first time he drove with it. It was too late for a test drive. Everybody was there—they just had to go.

He kissed Lisa, Eva, and Isla goodbye, and they were off...
Roen stood in the road, waving his hands to direct Zoltan
out into the street without demolishing any more community
property. The cameras were rolling... Zoltan, Maria, Roen,
and a handful of journalists made up the crew for this maiden
journey. Also traveling with them was the tour mascot, a $400
Meccano Meccanoid "robot," which Rachel had put together.
Nicknamed Jethro Knights, he could say a few words and move
his mechanical arms.

"It was so difficult to get out of my front yard," Zoltan recalls.
Maneuvering the 40-foot bus was not the same as maneuvering
a car. He ended up taking out a center divider on his street,
he says. He drove slowly at first to make sure the whole getup
didn't just fall apart.

"And here I am driving 55mph on the Golden Gate Bridge—
can you imagine if a 40-foot piece of wood flew off?" The coffin
top could smash a windscreen and kill someone.

But it didn't. No one needed to worry about the bus being
a safety hazard. As they drove along, in fact, they were getting
looks from fellow motorists that were the opposite of worry—
looks of wonder.

On the Golden Gate Bridge, with the windows open—the
windows had to be open because the air-conditioning system,
covered with the coffin-top accoutrements, was useless—it
started to feel like a proper road trip. The interior of the bus
wasn't luxurious, but it was built for comfort, like any RV. The
journalists were lounging on the bunks, laughing and eager
for the stories to come. The Transhumanist Party faithful
were seeing their hard work pay off. Zoltan was behind the
wheel and grinning—they were on their way. Everyone was
relaxed and excited at the same time. And every time they
were stopped in traffic, everyone else on the highway was
staring at the bus.

"And all of a sudden we realized this was going to be an enormous success."

***

Two hours later, it no longer looked that way. As the engine smoked, Zoltan was forced to stop. It looked like disaster.

The Immortality Bus was an incongruous name for something that was supposed to look like it belonged 6 feet under. But the project itself was a contradiction. It was a campaign about the future housed in a vehicle that was decades old, and showing its age. That contradiction threatened to trip them at the first hurdle.

The vehicle was severely overheated and there was "oil all over the damn road," recalls Zoltan. A problem seemingly from the Industrial Age had come to haunt them. They were still four hours from their first scheduled stop, in Tehachapi, California, where GrindFest was getting underway the next day.

As Zoltan looked at the oil trailing off into the distance behind them, and back to the still-smoking engine, panic set in. Then devastation.

If the engine was fucked, the Immortality Bus tour was dead. This was a 1978 bus; a replacement engine would not be easy to find. It could take weeks. It would probably be easier to get a new bus, and start over. The journalists would be long gone, and any chance Zoltan had of being considered a serious candidate would be gone too. He'd be a laughing stock—or worse, just totally forgotten. Did you hear the one about the presidential candidate who wanted to become a robot? He broke down before the election. The bad jokes wrote themselves.

Even if the engine could be repaired, how long would that take? They'd be off the road, when their first event—getting biohacked at GrindFest—was tomorrow. From GrindFest they were due to head straight on to the CTIA Super Mobility conference in Vegas, where Zoltan was speaking right before Mike Tyson. All that might have to be canceled, even if the bus could be made roadworthy in the coming days. Again, Zoltan would be the butt of jokes.

Just an hour before—after all the work put in, the relationships strained, to get this right—it had been paying off. Things had never been better. Now, the bus tour was teetering on a metaphorical cliff edge.

Zoltan rallied himself. The Immortality dream wasn't dead yet. When he'd faced storms on the boat, did he roll over? When they told him *Wager* was unpublishable, did he delete his manuscript? If there was any way at all they could keep going, he would find it.

He revved up the engine again and they drove back to the last town they'd come through, and pulled in at a gas station. He phoned his dad to get an old hand's advice—after all, his parents had once owned the same model Blue Bird—and learned what he already knew about the dire situation they were in if the engine was ruined.

On the plus side, the filmmakers were lapping all this up. It would make for a dramatic first act—if a way was found to continue the trip. Of course, if this was the end of the line, the movies would be short and comic.

Sometimes, you get lucky. Since Zoltan didn't believe in transhuman destiny or divine intervention, luck was all it was. After getting under the bus with a flashlight, and inspecting everything closely, he saw that there was a crack near the engine casing. This is why the bus was leaking oil, and why it was overheating. They couldn't keep going like this forever—the vehicle would continue to overheat and smoke, parts would melt, and it would stop running. But, if the engine was regularly fed with oil to replace what was trickling away, then they could keep going for a while. He made a stop at Rite Aid and bought $190 worth of motor oil.

And so Zoltan's merry band, his self-described "circus," in its metal coffin on wheels—an emblem of the future—continued its ride. They had to stop every hour to fill the engine with oil, which ran straight through and continued to stain the highway as they drove. At least now it was night, so no one else out on the road would notice the liquid leaking all over the place. However, it did continually splash onto the windscreen of one of their filmmakers, who was following behind in his own car.

They would make it to GrindFest in the Mojave Desert, despite some further hairy moments heading down a dirt road on a hill to where the event was happening—not to mention back up the hill again on the way out... They would make it to Vegas, where they put the finishing touches on the bus—the fake flowers on top—and found a more permanent solution to the oil problem. Zoltan bought a pan, like the kind used for a Thanksgiving turkey, at Walmart—and attached it to the bus to catch the leaking oil, so they could recycle it back through the engine rather than losing the oil to the road. Lisa and the kids would join them in Vegas too, for some family time and PR gold...

At 4 a.m. on that first night, Zoltan stopped the bus. He hadn't slept in three days, and now he was beginning to feel it. The adrenaline stopped pumping... They'd made it this far; they could make it further... They'd won their first big challenge.

The gang cracked open the beers and whiskey they had with them, and Zoltan produced a couple of joints. The stress melted as they drank and smoked through the wee hours... Right there in the 1970s Wanderlodge.

So it was that the Immortality Bus set forth, with a mission to bring transhumanism to the people...

Chapter 9

Transhumanism's Trump Card

Zoltan was about to become—for real—part-machine. He'd talked in public, and written in the media, for two years about how humans could benefit from merging their organic bodies with computer technology. Now, he was finally getting a microchip inside him. The RFID chip could be used to start his car or unlock his house with the wave of his hand. Or hold Bitcoin. When you stood nearby, it would send a text to your phone that read, "Win in 2016."

Zoltan was reclining in a dentist's chair in a makeshift lab at GrindFest, the biohacking festival in the California desert, where the Immortality Bus had finally arrived after its hairy first day on the road. Zoltan rested his right arm behind his head, while the chip was injected into his left hand, between his thumb and forefinger.

He'd experienced "much worse" pain over the last few days, he told the rolling cameras, mentioning that he'd fallen off his ladder while outfitting the bus. He thought he'd broken his finger at the time.

"It's one thing to be on Facebook talking about" getting a chip, he tells the cameras. "It's another thing to have these guys doing it." It was important for him to get the implant, not so much for the amazing things it could do, but because of what it represented. Modifying your own body with technology, helping make it acceptable and widespread, "is perhaps the most important thing you can do to move the movement forward."

GrindFest was full of folks who'd already "hacked" themselves, or wanted to. Folks who'd gotten magnets inserted into their fingertips so they could move small pieces of metal and experience the feel of electromagnetic fields. Or got chips in their arms that they could scan to provide them with real-time health data.

"I definitely consider these guys the forerunners of where the entire movement is going," Zoltan says.[1]

The hairy moments getting to and from GrindFest were far from the only ones in the months-long journey of the Immortality Bus across America. There would be that time the bus's wheels got stuck in an actual grave while they took a road through a historic cemetery in Tennessee. Or when they crossed a bridge that said "Max. 5,000 lb" and later discovered the bus weighed 14,000 lb—they could have died. Or that time they parked outside a crèche and the police arrived after receiving a call to say that someone had brought a giant coffin to a children's daycare center...

(The bus often encountered cops, Zoltan says—he was flagged down several times because the bus looked unsafe. But once the officers learned he was just passing through, that he wouldn't be in their state for long, they decided to let the bus go on its way, and let the next state worry about impounding it.)

These incidents would become smile-sparking memories, but they weren't what the trip was about. The purpose was to venture to as many places across the US as they could, do as many attention-grabbing events as possible, to bring a message of atheism, life extension, liberty and responsibility, and straight-up better living through awesome tech...

The day after getting chipped, Zoltan and crew were off to Las Vegas, where the massive CTIA Super Mobility tech convention was taking place and Zoltan was set to give a "huge speech." Well, he thought it was going to be huge. He was speaking just before Mike Tyson, so the hall began filling up as Tyson's slot got closer.

They pulled the bus, still leaking oil, into the parking lot of Vegas's famous Venetian Hotel. By the time they got there, they'd painted "Presidential Candidate" onto the side. The crew also used the Vegas stopover to pop the "coffin flowers" on top of the bus; to make another few fixes, and get the Immortality Bus website going.

Lisa, four-year-old Eva, and one-and-a-half-year-old Isla also joined Zoltan in Vegas. They all got to hang out with drones, rudimentary robots, and try on virtual reality (VR) headsets. Zoltan was in his element. It felt like the perfect environment for the would-be leader of an upgraded America: to be surrounded by all these cutting-edge commercial electronics. And it didn't hurt—from a publicity perspective—to have his photogenic family in tow. It was the most natural thing in the world for his four-year-old to be playing with a robot. A marker of the cool future in store for your kids. From a social-media standpoint, "it was a goldmine," Zoltan says.

After Vegas, Zoltan made a short stop in Newport Beach, where he gave a speech at his friend David Kekiche's yacht club. Kekiche, a wheelchair user, was a strong supporter of all Zoltan was fighting for. (Upon his death a few years later, he was cryonically frozen—hoping to one day come back in the shining transhumanist future that Zoltan and others would build.)

Following the stop in Newport Beach, the Immortality gang took a week off. "The bus trip wasn't a continuous thing," Zoltan points out. He would do a few stops, then go home for a bit, then pick it up again. Different journalists traveled on different legs and captured different moments.

But the success of that first leg at least ensured there was a steady stream of journalists showing interest in tagging along...

Zoltan needed wins like the big welcome he got at GrindFest, and the fun footage from the Super Mobility convention floor. Because a lot of the day-to-day slog on the bus was not fun. Driving a vehicle that could barely be maneuvered and might go on fire at any time; driving in extreme heat and without air con.

And, what's worse, all hell was about to break loose within the world of his fledgling Transhumanist Party...

***

Zoltan was doing all he was doing for the cause of transhumanism. And yet, he now found himself the subject of a major, sustained attack... Not from Christian fundamentalists, disability rights activists, or the usual suspects who had a problem with his views... But from fellow transhumanists.

After the Vegas leg of the bus tour, one of the embedded journalists, Dylan Matthews, published an article for *Vox* in which Zoltan was paraphrased as saying he planned to drop out of the presidential race and endorse Hillary Clinton, when the time was right. He hoped that, if he did this—and if he'd generated enough buzz as an independent candidate—he might be rewarded with some kind of position as a technology advisor to Clinton, whom he, along with most of the world, believed would win the presidency.

Transhumanists, supporters and opponents of Zoltan, went nuts. As Jamie Bartlett summarizes in his book *Radicals: Outsiders Changing the World*, this "throwaway statement" about endorsing Clinton "threatened to derail Zoltan's entire campaign."[2]

Nobody in the orbit of the Transhumanist Party had ever heard about this plan. It seemed like a major deal, something that should be discussed within the organization: who the party would endorse or not; whether it was right for the Transhumanist Party candidate to speak openly about endorsing a political rival. Serious stuff. Was the Transhumanist Party a serious political organization, or just a front for Zoltan's personal ambitions?

The Transhumanist Party had four "officers" when it was founded: Zoltan; his wife; Chris T. Armstrong, the hyper-fan who would go on to write a 500-page book about *Wager*; and another transhumanist friend, Hank Pellissier, a former managing director at the Institute for Ethics and Emerging Technologies. The most independent of the four, Pellissier, promptly resigned in the wake of the Clinton comment, describing it—in a lengthy web article about his resignation—as "the last straw." He wrote:

> The only meeting I attended [for the Transhumanist Party] was held via telephone. We followed the protocol required to set up a California non-profit. We nominated, seconded, and voted unanimously to have Zoltan represent our organization as a US Presidential candidate.
>
> I was happy, at the time, to do this. I considered Zoltan a friend...
>
> ...
>
> But...as Zoltan's campaign proceeded, I realized that he seemed to make all the decisions, entirely on his own, apparently. There's an entity called "The Transhumanist Party" but there weren't any strategy meetings involving me as an officer, or any group meetings, except one he hastily organized because a TV crew thought it would be interesting.[3]

Pellissier went on to criticize Zoltan's attacks on religion as counterproductive, his bus tour as a vanity project, and the Transhumanist Party as unrepresentative of the broad church of transhumanism.

Pellissier's piece helped give others within the community the push they needed to speak up in an organized way. The co-founder of the Mormon Transhumanist Association, Lincoln Cannon, started an online anti-Istvan petition. The opening line read: "We are transhumanists, and we disavow Zoltan Istvan's candidacy for President of the United States."

It went on: "We also disavow the nominal Transhumanist Party USA, so long as it cowers under authoritarian control, so long as it denies the diversity of Transhumanist values, and so long as it mongers unnecessary hostility toward others."[4]

Cannon wrote a separate open letter, making personal appeals to Zoltan supporters he called out by name, like Maria Konovalenko. Cannon's charge was as serious as it gets: that, because he had set transhumanism on a collision course with the current culture, Zoltan could literally start a war, exactly as it had played out in *The Transhumanist Wager*. Cannon compared Zoltan's brand of transhumanism to the Westboro Baptist Church's brand of Christianity, namely, offensive, evil, a perversion.[5]

Cannon was very much of the view that transhumanists needed to persuade religious people that transhumanism was compatible with religion. Since most folks were religious, transhumanism should work with religion as an ally, rather than against it. Cannon's approach was anathema to Zoltan.

Major names signed the petition, including Max More and Natasha Vita-More. More commented that Zoltan "is not representing transhumanism accurately or with good information," while Vita-More opined that transhumanism should be "inclusive."

Pellissier, of course, signed. And he followed up his resignation article with another, titled "15 Questions Zoltan Istvan Is Avoiding," where he asked, among other things: "Why do you frequently present yourself as a transhumanist who works harder than everyone else to extend life? Do you realize there are scientists solving the problems of aging, and you're not one of them? Have you actually raised any money for research, or convinced a single government representative to increase anti-aging funding?"

It was getting rather petty, now.

Pellissier did, however, confront Zoltan on a serious point in the "15 Questions" post: "You have stated that the Transhumanist Party is not really 'a party' but you collect campaign funds anyway. Isn't it illegal by FEC rules to pretend to be a political party, and ask for campaign funds?"[6]

Bartlett describes the backstory here:

> Before he declared he was running for president Zoltan tried to register the Transhumanist Party as a national political party, but he'd been unable to because he didn't meet the arduous Federal Electoral Commission (FEC) criteria. Not willing to give up, Zoltan registered a Political Action Committee (PAC) instead. A PAC is not a political party, but rather a simple vehicle that allows people to donate money to political campaigns. He cunningly called his PAC 'the Transhumanist Party'. He then registered himself as an independent candidate, running without a party. In short: Zoltan was running for president, but in the same way any lone individual can. But he could not do it as a candidate of the Transhumanist Party. This distinction is important. Under electoral law, it is forbidden to claim to belong to a political party if you do not; or to receive donations as a representative of a political party that does not exist. Zoltan had been doing both.[7]

Zoltan's response to his critics, quoted in Bartlett's book, was typically pugnacious: "I'm trying to set myself up as someone who has broken laws for the benefit of Americans. I want to openly break them. I'm ready to be arrested. It's good press."[8]

Zoltan makes the same point to me. His critics were looking at what he was trying to accomplish in entirely the wrong way. FEC recognition and obeying the technicalities of electoral law were "completely irrelevant": "Every revolution that's ever started in a country did not start because it was legal."

There were prominent voices, for sure, who spoke up in Zoltan's defense. Giulio Prisco wrote that the Transhumanist Party shouldn't be taken seriously in terms of electoral prospects, since it was a "single-issue fringe party." It was a "daring publicity stunt," and Zoltan should be commended for garnering media attention where others couldn't. He added, to counter the idea that Zoltan was inciting violence: "I am persuaded that Zoltan's militant atheist bigotry is mostly posturing to grab media attention."[9]

Cannon also admitted to me that he started the petition, at least in part, to try to generate publicity. They were all, in their own way, attempting to bring attention to transhumanism.

Still, the critics wanted a political home for transhumanism that didn't include Zoltan. The petition to disavow quoted the Transhumanist Declaration, a statement of transhumanist beliefs crafted by more than 20 authors from around the world, which had come to be broadly accepted as foundational principles, and was adopted by the board of the largest transhumanist organization, Humanity+, in 2009. This was an example of the kind of collaboration and cooperation, bringing everyone together, that was possible within the movement— when no one acted like a dictator. Zoltan's critics went about trying to form a political organization that explicitly adopted the Transhumanist Declaration, rather than the Transhumanist Party platform, or Zoltan's forthcoming Bill of Rights. They called it the Transhuman National Committee.

All this was the last thing Zoltan needed as he plowed on with his bus tour. But it was good for him in at least one respect: It proved, to him at least, that he had been right all along. This agonizing about the way he'd won over the media, or arguing about the minutiae of his party setup, demonstrated exactly what he'd always thought was wrong with the institutional transhumanist movement: obsessed with its own workings; not looking outward to create a big buzz and become popular.

Zoltan was determined that the Transhumanist Party wouldn't become a talking shop on his watch; another third party in American politics concerned with the granular detail of philosophies and policies that would never get put into practice. The Transhumanist Party was not going to be another forum for transhumanists to talk to themselves. There were enough of those.

The transhumanist movement, in Zoltan's view, had long been too esoteric. These guys—and most of them were guys— were geeks. Having spent, by this point, a huge amount of time in the movement, he was convinced that the majority of people in it had various social and mental issues—"I might be one of them," he quips. They had social anxiety; they were loners; they were obsessed with their own logic and lacked empathy; they often didn't have families—whether it was because they were too young or just had no interest... A lot of transhumanists, in his experience, were gender queer—which made sense since, at its core, transhumanism is about transcending set identity. So far, all so good.

But... Since transhumanism attracted people outside the mainstream, it was staying a fringe movement. Conquering death was too important an issue to remain on the fringe—everyone should be demanding immortality. That meant transhumanism had to break into the mainstream and get massive mainstream media coverage.

The finer details of doctrinal disputes between Zoltan and other transhumanists could bore the voting public to tears. It isn't something most of us would likely care about: the difference between the Transhumanist Declaration and Zoltan's Transhumanist Bill of Rights, for example.

The deeper essence of the conflict was this: Certain veterans who'd been around the movement for a long time felt that transhumanism had a legacy that should be respected and a future that should be nurtured in accordance with the legacy. Zoltan just wanted to beam the word "transhumanism" and the idea of living forever through science into every home in the world, and he didn't care what hare-brained schemes he had to plan, or what controversy he stirred up, to do it. He wasn't steeped in the history of the movement—he didn't care about the history; he didn't know it. Transhumanism was a cool idea he'd found and he'd decided to be its champion. He wanted to live forever—and wanted everyone else to be able to live forever, too. The notion that he was part of a tradition that he should be safeguarding was not in his programming.

Transhumanists had spent decades refining their beliefs, and this had resulted in the Transhumanist Declaration (1998), which has been updated through the years. Stating that "humanity's potential is still mostly unrealized," the document declares: "We envision the possibility of broadening human potential by overcoming aging, cognitive shortcomings, involuntary suffering, and our confinement to planet Earth." It calls for careful deliberation on "how best to reduce risks and expedite beneficial applications" of advancing science and technology, and for "forums where people can constructively discuss what should be done." It is important to show "solidarity with and concern for the interests and dignity of all people around the globe." There's even a clause in support of the wellbeing of animals.[10]

This 'Kumbaya' stuff and deference to consensus-building was not found anywhere in Zoltan's Transhumanist Party platform, or in how he conducted himself on behalf of transhumanism.

People were pissed off that Zoltan didn't discuss his platform with elders in the movement, and that he didn't pay respects to and adopt the Transhumanist Declaration in his party principles. He had made no effort to reach out to and bring on board legacy transhumanist groups, like Humanity+, before he set up the Transhumanist Party. He had simply decided to do his own thing.

He did it all off his own bat: registered the trademark for the Transhumanist Party, set up bank accounts, an LLC. Set up a website. Then announced he wanted to be the presidential candidate of the party he'd just founded. Then he got his friends to vote for him as the organization's candidate.

Of course, in the end, Zoltan couldn't technically run as the Transhumanist Party candidate for president, since the Transhumanist Party wasn't officially recognized as a political party by the FEC. He was running as an independent candidate, who was a transhumanist and founder of the Transhumanist Party PAC, endorsed by its officers as someone they'd like to see become president. But that's a distinction without a difference for all but the most ardent political anoraks.

As far as Zoltan was concerned, he was really the only hope the transhumanist community had to make a splash in a presidential campaign.

Perhaps someone else within the movement would have been better at uniting the different factions around a common manifesto and making a serious, philosophical, and practical case for transhumanism in politics. That person would have received a tiny fraction of the interest that Zoltan's campaign created. The media aren't interested in boring centrism from minuscule fringe parties. That ground is already well occupied by the major players. Candidates with microscopic support only make headlines when they do batshit crazy stuff. And Zoltan wanted to generate headlines with "transhumanist presidential candidate" in them...

Even though Zoltan's preferred candidate for the presidency, were he to drop out of the race, was Hillary Clinton (mainly because he agreed with Democrats over Republicans on redline social issues like abortion and LGBT rights), in many ways Zoltan's candidacy more closely resembled that of Clinton's Republican rival. And not just because they were both real estate guys.

"There's two different ways to win the presidency," Zoltan tells me. The first is building up a political career from small beginnings, running for local office, working your way to state level, maybe becoming a governor or senator; gaining enough supporters and allies along the way that you have the chops to run for president. "Or you could be like Trump, and be such a loudmouth and so well known, and such a rich person, that you have a natural advantage."

Zoltan chose the latter route—obviously.

The transhumanist movement needs people who are "Trump-like," Zoltan says; who understand how to create spectacle and are "just totally colorful, totally eccentric, totally out there."

Zoltan didn't exactly model his persona and candidacy on Donald Trump's. Zoltan was already generating publicity by attacking the status quo before Trump got into the 2016 race. But they both had a flair for extremes that were sure to piss people off and get everyone talking. Trump wanted to build a wall to keep Mexicans out. Zoltan suggested opening the borders, but monitoring the immigrants with drones and microchips. Trump wanted to ban Muslims from entering the US. Zoltan proposed canceling disabled people.

Both were despised and attacked by the establishment wing of their respective movements. Zoltan was condemned for being too outlandish, divisive, and controversial a figure to represent transhumanism. The same thing happened to Donald Trump in 2015/16: Fellow Republicans lambasted him as unfit to represent their party. He was accused by former Florida governor Jeb Bush of not having the "temperament or strength of character" to be president.[11] Senator Lindsay Graham said Trump was a "race-baiting, xenophobic, religious bigot." Senator Ted Cruz called Trump "utterly amoral."[12]

Zoltan never voted for Trump. And yet, throughout our interviews, it's pretty clear from the way he talks that Zoltan admires Trump. At least, admires what he accomplished as a candidate—his "Fuck you!" to the establishment. The fact that, for all his bizarre stunts and supposedly grotesque views,

Trump won on his own terms, turning American politics upside down.

Zoltan longed to accomplish something similar. Not to have to put in the hard graft of rising through the ranks in a local party, putting in time as an alderman or a state senator or something meaningless before aspiring to national office. But to play on his success in media and business, and swoop to victory by tapping into some vein of mass support everyone else had overlooked.

***

Zoltan called himself "The Science Candidate" on social media. But "science" conjures up images of sobriety and methodology. Zoltan's journey was more like a crazed rock tour mixed with guerrilla politics.

The Meccano toy robot, nicknamed Jethro, rode shotgun while Zoltan drove the Wanderlodge cross-country. Zoltan brought a big box of cassette tapes, filled with all kinds of music, which he played over and over again as they rode. The passengers lounged in the back on the 1970s beige and orange built-in beds and sofas. The windows were always open, since the air conditioning didn't work. Zoltan's promise to fellow travelers on the trip, journalists and others, was a heady one: Come with us and you can get chipped, hang out with robots, experience virtual realities, get on the inside of life-extension cults, meet folks at the bleeding edge of crazy and the frontiers of what's possible.

The trip proved a strain on family life, no doubt about that. Lisa was at home with two young kids while her husband pursued his fantasy. "First of all," Lisa tells me, the bus is "obviously not safe. Who knows where and when it could break down." She was constantly faced with the thought: "Oh, my husband's going to die in his coffin bus."[13]

There were other stresses, too. Zoltan was spending their money on his campaign against death in this ancient brown bus, and they now also had to pay out hundreds of dollars a week for a nanny, with Zoltan off gallivanting and Lisa working. Once, while Zoltan was on the road, the toilet broke at home in Mill Valley, and the stress for both husband and wife just mounted as they went back and forth on the phone. When his daughters came down with a bug, Zoltan was holding a placard promoting himself outside the Democratic National Convention. It wasn't the best advertisement for his "good dad" skills. His guilt is captured in Daniel Sollinger's documentary about the bus tour, *Immortality or Bust*. Zoltan was also still trying to run his vacation-rental business from the road, do his job as a landlord, dealing with tenants' hassles. And his father—who'd had several heart attacks by this point and lost an eye—was becoming sicker again... and might not have long left. Zoltan had to tune out a lot so that he could focus on this wild, doomed presidential bid.

Apart from all that, just in terms of the practical business of driving the bus and getting around, it wasn't easy. Driving the bus was "really tough," Zoltan says. The Blue Bird was supposed to have power steering, but that didn't work. Because they had to run with the windows open all the time, smog and smells billowed into the bus in every city they went through. Driving up the hill to get out of GrindFest, Zoltan had everyone get out of the vehicle because he was afraid the whole thing would just topple over. Every day of the three-and-a-half months they were on the road, he wondered if this would be the day the bus gave out on them.

The Immortality crew were a curiosity wherever they went. When they pulled into parking lots, folks would approach, ask what it was all about... Zoltan handed out business cards, told them he was running for president, what he stood for. Most folks were more bemused than enthused. But Zoltan was reaching the people, taking transhumanism directly to them. More folks probably learned what the word meant than ever wanted to.

The bus crew took a week off post-Vegas, and then came back together to enjoy a fundraiser for their mission, hosted by a friend of Zoltan's at his house near Laguna Beach. From there, the bus trundled on to San Diego, and Zoltan and six Transhumanist Party supporters stood on the pavement in front of the USS *Midway* aircraft carrier, holding signs protesting the end of the world. "Give NASA Funding for Asteroid Protection"; "AI Must Be Friendly"; "Make Love, Not Viruses"; "Prevent a Global Pandemic!" the signs read. Zoltan held the kicker: "Transhumanist Party Prevents Existential Risk."

Dmitri, the Moldovan who had substantially funded the tour with his big donation to the Indiegogo campaign, had a lieutenant at his existential risk think-tank join the bus for this leg. Alexey, who was himself a passionate believer in the power of science to extend life, had frozen his mother's head after her death in the hope of one day reuniting with her.

On October 14, the gang took a tour of the Alcor cryonics facility in Scottsdale, Arizona, which brought some attention from local TV news. CEO Max More described the workings of the facility's giant metal canisters, each containing four full frozen bodies plus five additional frozen heads. The hope was that one day all these dead people could be brought back to life. "Cryonics is simply an extension of emergency medicine," More explained.[14] Folks had paid $200,000 to have their bodies kept like this. It was only $80,000 if you just wanted your head preserved on the assumption that in the future they would be able to fashion a new body for you. At the meeting, there was no warmth between the two transhumanist titans, More and Zoltan. Not surprising, given that More had recently signed the petition to disavow Zoltan as a transhumanist presidential candidate.

Zoltan has been asked many times by journalists to give his opinion on when cryonics might actually catch on, become sought after. "It'll probably catch on only when they bring someone back to life," he laughs. Makes sense. If they ever do revive a person who's been cryonically preserved, everyone will want in. Until then, it's just another madcap thing some rich people do.

The interest and goodwill from local news crews was not unique to the stopover at Alcor. When you show up in folks' hometown, Zoltan says, "and they can come aboard and they meet you, and they realize you're a nice human being..." Zoltan truly believed there was a chance he could persuade even the most ardent opponents of life extension and transhumanism by meeting them, having a conversation, laying out his logic. Not everyone would respond positively, of course, but he believed he was winning hearts and minds, and that there was no more urgent struggle.

The crew tried to make a point of doing something in every state they passed through, Zoltan says, "and something different, too."

In Alabama, Harlem futurist and hip-hop artist Maitreya One met them, and they went to the Greyhound bus station that houses the Freedom Riders museum, celebrating those who protested segregated bus terminals in the South in the 1960s. "Maitreya is a civil rights link from the past to the future—and one of the few African-American transhumanists I know," Zoltan wrote in a *Newsweek* column about this stopover. "Like others in the burgeoning transhumanism movement, Maitreya supports becoming a cyborg in the future, and he knows the coming controversy over such aims may end up as challenging as the civil rights era battles over racism."[15] Zoltan tried to make a link between those who rode buses to protest racism in the 1960s and his own bus journey, arguing that the fight to modify your own body however you saw fit, to become a cyborg, and to never die, was the civil rights struggle of the twenty-first century.

Also in Alabama, the Immortality Bus paid a visit to one of the Bible Belt's megachurches, Church of the Highlands, with a congregation of 32,000. The gang pulled into the parking lot of the university-sized campus, and slunk inside, cameras and all. Journalist Veit Medick from *Der Spiegel* was also with them at that point.

At first, it was going OK. They grabbed a brew at the campus coffee shop. Wandered around the massive, empty church auditorium. "This is amazing," Zoltan said out loud to himself, staring at row after row of seats intended for thousands of worshippers. A passing pastor, who noticed the new faces, offered to show them around. Zoltan started a conversation with him about the nature of the future. They talked about using virtual reality headsets to teach about God. They talked about whether the souls of robots, AI, or aliens could be saved by Jesus.

"It's really the first time I've thought about whether robots or artificial intelligence could be saved," the pastor, Kyle Cantrell, said, "but it's an interesting concept."[16]

Nobody from the church had noticed, yet, the Immortality Bus sitting in the parking lot. Nobody knew that's where Zoltan & Co. had come from.

Then word got out: There was a massive coffin-on-wheels on church grounds mocking the churchgoers and their faith. It had "Science vs. The Coffin" written on the back, and the text on the side seemed to suggest that "Transhumanist Zoltan Istvan" could provide "Immortality," rather than Jesus. Somebody probably googled Zoltan and learned he was a kind of atheist guerrilla performance artist...

All of a sudden, a message rang out on the campus PA system—the church had been infiltrated. Stay where you are. The campus was on lockdown.

"We were there about 90 minutes before they got us," Zoltan says.

Plain-clothed men with guns approached. "In Alabama, everyone has guns," Zoltan points out.

The bus crew were instructed to leave—and escorted out by armed men.

"Anyways, they were very nice people," Zoltan says of the gun-toting church folk. "They weren't mean in any way. They just said, 'You didn't come with an invitation, and you're on private property.'"

The bus trundled on to its next stop...

The awe Zoltan had expressed in the church auditorium was genuine. And mixed with more than a tinge of jealousy. How can one single Christian church boast 32,000 members, "when I can't even get together 100 people at an event"? How was Christianity so successful at selling itself? Why did people want to believe in an afterlife, rather than transhumanism's eternal life here on Earth?

It was comforting to remember that Christianity had had a tough time starting out. It was hundreds of years after the death of Christ before the religion named for him became widely adopted and accepted as a moral way to live. But Zoltan couldn't afford to wait centuries for transhumanism to become the world's most adopted philosophy... He needed eternal life before his inevitable death.

The visit to the megachurch did help reinforce, for Zoltan, the reasons why Christianity—or any religion—was easier to sell to the masses than a transhumanist philosophy. Religions were selling an idea that could never be disproven: Paradise awaits after death; all you have to do to get there is behave in the right way. Transhumanist eternal life, however, depends on a lot more than belief. It depends upon constant work every day by scientists and inventors and activists demanding change, upon continual progress. And it can be disproven. Science will either create immortality here on Earth, or it won't. But, if physical immortality proves impossible, religions will still be able to sell spiritual immortality in an undiscovered country. Progress is hard; it requires a lot of actual work. Faith is easy; it's just a thought process.

The wide disparity between the numbers Christian megachurches could attract, and the numbers transhumanist-type ideas could attract, was on full display when the Immortality Bus headed south from Alabama to visit two "transhuman spiritualist" communities in Florida.

Although well-attended for Zoltan's speeches—both locations hosted around 100 people while the candidate was there—it was a far cry from crowds in the tens of thousands.

Zoltan spoke at a service at the Church of Perpetual Life near Miami, a community devoted to assisting "all people in the radical extension of healthy human life," and "providing fellowship for longevity enthusiasts through regular, holiday and memorial services," according to its website.[17] The Church did not care whether you were an atheist, agnostic, or believed in divinity. Members were simply united in their desire to live very, very long and healthy lives through science.

Zoltan gave his standard "stump speech"—a version of which he gave wherever he formally spoke along the tour—explaining the importance of transhumanism, his campaign, and the bus, with the aid of PowerPoint. He received a big round of applause and stayed on stage to answer audience questions.

One man asked perhaps the most frequent question Zoltan received, wherever he gave his speech: If we're all living forever, what about overpopulation? The candidate gave his standard response: We should focus on geoengineering the planet to support more life, rather than sacrificing ourselves. (With the ability to live in the oceans, in space, and in virtual realities—because we've re-engineered the physical universe and our own minds and bodies—supporting billions more beings won't be a problem.)

After Zoltan's talk, Church founder Bill Faloon (who made his fortune selling dietary supplements) spoke about what you can and should do to combat high blood pressure. "Our church is all for any technology that will enable human beings to escape the confines of nature," he told the crowd.[18] Working with your doctor to get your systolic below 120—taking meds if necessary—was just one way to keep a natural death at bay. Faloon also opined on the benefits of diabetes drug metformin for lengthening your life.

In Orlando, Zoltan was hosted for lunch at Terasem, perhaps the standard-bearer for transhuman spiritualism. The name comes from the Greek, "Tera" for Earth and "Sem" meaning Seed; Earthseed is a religion in the science fiction of Octavia E. Butler, which inspired Terasem's founders, Martine and Bina Rothblatt.

The couple wanted "to find a way for people to believe in God consistent with science and technology so people would have faith in the future," Martine, a transgender media and biotech millionaire, once told *Time* magazine.[19] That article explained Terasem's philosophy:

> Organized around four core tenets—"life is purposeful, death is optional, God is technological and love is essential"—Terasem is a "transreligion," meaning that you don't have to give up being Christian or Jewish or Muslim to join. In fact, many believers embrace traditional positions held by mainstream religions—including the omnipotence of God and the existence of an afterlife—but say these are made possible by increasing advancements in science and technology.[20]

At Terasem's ashram, which sits right on the ocean, computer servers host thousands of "mindfiles," thoughts and memories recorded in video, audio, and text, and sent to Terasem. The idea is that, one day, these mindfiles could be used to recreate a person's consciousness. Two onsite satellite dishes beam mindfiles into space in the hope that these fragments of humanity will reach aliens. If humanity never reaches a point where we're able to bring the dead back to life, maybe aliens will be able to resurrect you from your memories.

Zoltan once again gave his standard stump speech at Terasem. But this time there was a twist: He gave the speech in the virtual world of videogame *Second Life*, several days after he'd actually left the Florida venue. Terasem created avatars for him and other speakers at their Annual Colloquium on the Law of Futuristic Persons, which was set up to take place within *Second Life*.

"A lot of the transhumanist community spends a lot of time in virtual reality," Zoltan says. He was one of the few, by his estimate, who wasn't that into videogames or online worlds. In fact, Zoltan had never experienced *Second Life* before his "campaign stop" there. He was really impressed with how like him his *Second Life* avatar looked, with its mop of blond hair and stubble. "I can never shave too closely," he points out, so they got that aspect of his physicality down. Even Zoltan's PowerPoint was ported over into *Second Life*.

Zoltan was campaigning in the "metaverse" before the "metaverse" was even a thing.

He was meeting potential transhumanist supporters wherever they happened to be—even in virtual reality.

***

After visits to biohacker festivals, cryonics labs, transhumanist churches, and even crossing the border into Mexico for a day (to take photos with a sugar skull)... After many nights in cheap motels; evenings spent unwinding at casino tables... All the worrying moments when Zoltan thought the bus wouldn't start... Coverage by journalists from the *New York Times*, *The Verge*, *Der Spiegel*, *Business Insider*, *Playboy*, and more...

After encounters with strange and controversial cult characters... Like John McAfee, the eccentric multimillionaire cybersecurity maven. McAfee was also running for president as the candidate of a self-started party, the Cyber Party. Journalist Anthony Cuthbertson arranged for the pair to meet for dinner at a hotel in North Carolina. McAfee's wife, Janice, was also there. The next day, the McAfees took a tour of the Immortality Bus.

"I can't think of a more horrific concept than immortality," was John McAfee's assessment of Zoltan's aims, as reported by Cuthbertson in the *International Business Times*. "It is anti-evolutionary," McAfee explained. "We need to die and die young preferably; dying is the most beautiful of all things. I'd get behind a platform where you kill everyone at 30."[21]

McAfee did think, however, that sex with robots was worth exploring.

Another strange encounter was Zoltan's appearance on Alex Jones's *InfoWars* YouTube show, after being repeatedly bashed by him. Jones, the notorious promoter of conspiracy theories about a secret, totalitarian world government involving the Freemasons, referred to transhumanists and campaigners like Zoltan as "porch masons": "They're not on the inside but they know what's going on, and they want access to what the establishment is building in its giant New World Order plan that's about a hundred times bigger than the Manhattan Project."[22] The New World Order would undoubtedly want some of the things Zoltan advocated, like mass drone surveillance and editing out human "flaws" at the genetic level... Zoltan was driven to the studio at an undisclosed location and interviewed by Jones's stand-in, David Knight.

After all that... After Zoltan churned out a ton of his own articles about the tour, for the major sites where he wrote columns... After the Church of Perpetual Life and Terasem... the Immortality Bus finally headed north toward Washington, DC.

Zoltan hadn't yet written the Bill of Rights, the transhumanist "95 theses" he planned to nail to the wall of the US Capitol, the symbol of the out-of-sync Old World, hoping to start his own Reformation. This event had been the anticipated culmination of the Immortality Bus tour since he first sketched out his route. But it was only as they left Florida that it hit Zoltan—he needed to start thinking about and actually writing the thing.

They rolled into DC in mid-December, 2015. The old Blue Bird was now firmly on its last legs. The engine was smoking; the tires looked like they were coming apart. The bus didn't start every time he turned the keys. There was this godawful dinging sound related to the air pressure in the brakes; the ding would just start randomly for 10 or 20 minutes at a time.

Zoltan decided not to take the bus into downtown DC right away, in case they fell foul of some vehicle standards and got busted. The Capitol and its surrounds were swarming with security. The coffin bus was a big red flag, and their mission already needed to be clandestine. Another guerrilla operation. Zoltan and his merry band parked at a hotel on the outskirts and headed into the city by foot and taxi to scout out their mission...

On December 13, sitting on the steps of the US Supreme Court, Zoltan typed out the Transhumanist Bill of Rights on his laptop. It was one of those times when his brain was right in the zone—and the content just flowed out of his mind onto the virtual page. The final document was only seven paragraphs long—a preamble plus six articles—and only took about 20 minutes to complete.

The preamble declared: "Whereas science and technology are now radically changing human beings and may also create future forms of advanced sapient and sentient life, transhumanists establish this TRANSHUMANIST BILL OF

RIGHTS to help guide and enact sensible policies in the pursuit of life, liberty, security of person, and happiness."[23]

The Bill declared that humans, AI, and cyborgs are entitled to indefinite lifespans; that hindering life-extension science should be a crime; that sapient beings are entitled to modify their bodies however they see fit; that mitigating existential risk is a guiding principle; that space should be explored in order to secure a future beyond Earth; and that "involuntary aging" should be classified as a disease.

Zoltan took the laptop to a printing place and printed off the single page of text.

The next day, they risked driving the bus in... since it was also part of their publicity tour, and this was the last day. They were meeting other Transhumanist Party members in DC. They pulled into the Capitol parking lot.

Zoltan—followed by Roen and Daniel with their cameras, and a handful of party members—marched to the US Congress armed with his "6 theses" and some sticky tape to affix his revolutionary document to the side of the building, where any passerby could stop to read it. Zoltan made sure to wear a nice shirt and suit pants that day, so he looked like a professional who was meant to be there, not a protestor who might get shot. His skin color was a privilege that protected him like a little bit of armor too. They'd carefully chosen a weekday, Monday, for their mission, when the Capitol wasn't swarming with visitors and they could actually get close...

At the Capitol steps, cameras in tow, Zoltan made a move to post the Bill on the front of the building. He was immediately stopped by an armed officer.

"You can't do that here."

Zoltan started to argue, politely: "Can we give it to you?"

"No."

Foiled, the transhumanists changed tactics...

Zoltan could still get the footage he needed for posterity and publicity. Like some kind of next-gen town crier, he stood in front of the US Capitol and read aloud his theses while the cameras rolled. The passing crowds glanced at him with curiosity and skepticism... Just another weirdo protestor with nothing better to do than stand outside Congress yelling about the problems in the world.

When Zoltan was done, the transhumanists tried again to leave their mark on the Capitol building, this time scuttling round the side... Zoltan casually ran up the steps when he didn't spot any police around, and managed to fix the document to the marble façade—but only for a moment. The page wouldn't stay stuck, kept falling to the ground. The sticky tape was no good.

As Zoltan tried to make it stick, he and the group were surrounded by officers with machine guns... "It got really tense really quickly," he says.

Zoltan had always argued that the cause of transhumanism might require acts of civil disobedience. It might be worth becoming violent. It might be worth getting arrested.

Was it worth getting shot?

From when they first took the Immortality Bus out from his home in Mill Valley three months before... Through all the worries that it might break down... The biohacker and technology shows... The campus talks... Protesting existential risk and nuclear war... Meeting and trying to convince ordinary voters, as well as fellow travelers outside the mainstream like John McAfee... To the genuinely scary moments like the time at the Alabama megachurch, or right now at the Capitol... When it came to media excitement, attention on his campaign, and spreading his name far and wide, it was hard to argue that the bus tour had been anything but a success.

Zoltan didn't need to become a martyr. He was already a media star. He didn't want to add to his arrest record and make winning an election even more difficult down the line. He didn't want to endanger the young believers who had traveled with him round America, any more than he already had… Nothing good would come from pressing the issue. They had done what they needed with the Bill of Rights, he says: "And had officially kind of posted it, even if it fell off."

The merry crew from the Immortality Bus skulked away from Congress… Zoltan had got what he wanted.

"We got lucky too," he concludes, "because it was raining later."

Chapter 10

An Eternal Campaign

Let's take a minute to consider an alternate reality. As Zoltan puts it: "When you're a year into the elections, people believe anything." Political faithful of all stripes start to imagine they're destined to win, no matter the polls, and the media run stories on the possibility of all sorts of election outcomes.

There's a very interesting scenario that happens in a presidential contest in the United States, if no candidate achieves 270 Electoral College votes. It's a scenario that Zoltan, for a brief time, thought could happen—and thought he might benefit from... He might even have had the backing of a major political party in the process.

The Immortality Bus generated so much media coverage, and Zoltan's campaign created so much online buzz—particularly among young people—that even the political establishment took notice.

Things took off in the wake of the bus tour. On that rainy day in DC after the drama on the steps of the Capitol, Zoltan dropped off the Transhumanist Bill of Rights at the office of his senator, California's Barbara Boxer. This didn't have the drama of sticking the document, guerrilla-style, to the building itself. But it was a way to deliver it officially and say it had been delivered officially to the United States Congress. The earlier stunt was for the cameras.

The Immortality Bus was left in long-term parking in Virginia, and the Bill of Rights was published and shared widely. Zoltan did a piece on it for the *International Business Times* as well as the *Huffington Post*, and others picked it up from there.

Even after the bus tour had ended, the media coverage of it led to more coverage and constant interest in Zoltan and his quixotic campaign. He was in demand for interviews and events all the time. He went to the 2016 Republican National Convention in Cleveland and the Democratic National Convention (DNC) in Philadelphia and was interviewed as a fellow candidate of Clinton and Trump. In an on-camera segment for The Young Turks, which has a huge following, Zoltan was introduced as the person "currently running fifth for president of the United States," even if his views about fixing climate change through geoengineering and solving biodiversity crises with cloning were treated with amused skepticism.[1]

At both conventions, what Zoltan was most proud of was that he and a handful of supporters spent much of their time counter-protesting Christians. One photo from outside the DNC shows Zoltan holding a poster up in front of a group of extremists with megaphones and placards. Zoltan's poster reads, "The Transhumanist Party Builds the Future," while signs hovering over his head warn that "HOMO SEX IS SIN" and "JUDGMENT IS COMING!" Zoltan wrote for the *Daily Dot* that it was important for atheists and transhumanists to confront "religious deception," because otherwise Americans could be "living in a Christian nation indefinitely."[2]

Zoltan summed up his convention experience in another article for *Vice*: "Somewhere between a roaming white llama, a purple face-painted dancing mystic, and a pack of born-again, sign-waving Christians screaming that I was going to burn in hell, I saw the irritated soul of America."[3]

The controversy and the attention Zoltan was generating clearly paid dividends when it came to "brand recognition" as a candidate and opportunities to get his message out on a national stage. Google data showed he was among the most-searched third-party candidates; in fact, the fourth most-searched overall, behind only well-known names like John McAfee, Libertarian Gary Johnson, and the Green Party's Jill Stein.[4] This emboldened Zoltan to make the claim, on social, in interviews, and when he met voters, that he was "running fifth" in a potentially knife-edge election. He was included, along with Johnson, Stein, and three others, in an online text debate hosted by I Side With, the viral questionnaire that matches voters to candidates. Post-election, *Newsweek* ran a feature (by Anthony Cuthbertson) saying the votes going to "third-party candidates like Jill Stein, Gary Johnson and Zoltan Istvan" rather than to the two major parties had likely affected the result.[5] Zoltan was a contender. Not to win, but to make a difference in who did.

Or... Was there a path to actually winning?

In the heady days of the campaign, it seemed like the impossible had a chance. The Democrats and the Republicans had decided on two nominees who were colossally unpopular with most Americans. Both parties' candidates had been in the public eye for decades, and faced scandals and questions over their character. Voters wanted alternatives. Already, Senator Bernie Sanders, an Independent, had come close to dethroning the presumptive Democratic nominee. The Republican nominee had never worked in politics before, and his breakthrough surprised everyone who was supposed to know better. The recent success of anti-establishment "Pirate" parties in Europe had proven that unknowns could emerge and win elections... What about in the race to be Leader of the Free World? Did a third-party candidate have a shot in 2016?

The favorite to become the presidential nominee of America's third-largest party, the Libertarians, thought so. And Gary Johnson began to take notice of Zoltan.

Here's the key point: The way America's electoral system works, a challenger candidate would not have to win very many votes nationally. All a third-party candidate had to do was win one or two states, and prevent both Trump and Clinton from getting to 270 in the Electoral College. Then, it didn't matter who was ahead, either in the Electoral College or the popular vote. If no candidate reaches 270, the US House of Representatives decides who becomes president, and the top three candidates in the Electoral College race can be considered. Each state's House delegation gets one vote, and the presidential candidate with a simple majority of the states, at least 26 votes—after multiple ballots if necessary—wins. In this scenario, the vice president is elected by the US Senate: The candidate with at least 51 senators' votes wins.

Johnson was a former two-term Republican governor of New Mexico. He'd been the Libertarians' candidate for US president in 2012, and gotten the highest number of votes ever received by a Libertarian in a presidential contest. He was set to be nominated again in 2016. Officially the party chose the presidential and vice-presidential candidates separately, but in practice they were likely to go with whomever Johnson recommended for VP. And Johnson considered Zoltan as a choice. According to Johnson, Zoltan asked if he could make a pitch to become Johnson's running mate, and Johnson agreed to meet.

Johnson was an Ayn Rand fan, and he knew that Zoltan had written a book that was compared to her work—that Zoltan had portrayed himself as something of an Ayn Rand figure for the info-tech age. Johnson had once given his fiancée a copy of *Atlas Shrugged* and said: "If you want to understand me, read this."[6] Zoltan was generating impressive traffic, and his ideas were intriguing. Johnson—like most politicians—was looking for a way to attract voters beyond his base, especially young voters. Zoltan was young for a presidential candidate, with a photogenic family. He spoke the language of the technotopians who were everywhere in the culture; and he had a youthful following. In the run-up to the 2016 Libertarian Party National Convention, Zoltan was invited to spend a day with Johnson at his home in New Mexico.

Never one to miss a publicity opportunity, Zoltan emailed journalists to tell them he was interviewing with a major candidate for their VP slot: "I don't want to say more, out of respect for the presidential candidate and their privacy," he wrote in the email.[7] At the time, Johnson was polling at 11% nationally, so Zoltan felt justified describing him as a major candidate. But of course, ambiguity was the goal, and it paid off. Two articles came out—in *CNET* and *Business Insider*—speculating on whether Zoltan was interviewing with Clinton, Trump, or Sanders. Johnson wasn't mentioned. "The Sanders, Hillary Clinton and Donald Trump campaigns did not immediately respond to requests for comment," Eric Mack wrote for *CNET*.[8]

Zoltan sighs when I ask him about meeting with Johnson. "I can tell the truth now," he laughs. He was really excited for it. He flew to New Mexico and rented a car, drove several hours to get there.

Within "three minutes" of being greeted by Johnson at his house, shaking hands, and entering, Zoltan says, "We go to his stove, and he pulls out a big joint. And he lights up and he says, 'Let us smoke.' Dude, it's three minutes. It's three minutes... And I got so high," he giggles. Johnson had been CEO of a medical cannabis company since 2014, until he stepped down to run for president.

It was just the two of them in this big house. They smoked, Johnson cooked seafood pasta, they watched *Orphan Black*—a sci-fi show about clones—and talked about how they could join forces to capture the votes of disillusioned boomers and radical millennials, pushing the presidential race to the US Congress. If they could win just one or two states, and Congress got to decide who became the next president and vice president... Might some establishment Republicans vote for a former two-term Republican governor, Gary Johnson, as a "safer pair of hands," rather than their own party's unpredictable official nominee, Donald Trump? Might some Democrats vote for

Johnson for the same reasons, if Republicans held the majority, and the alternative was President Trump?

This was the dream scenario that was deeply considered that night in New Mexico by the Libertarian and Transhumanist Party candidates, amid weed and sci-fi.

***

In the end it wasn't to be. Zoltan and Johnson got on well, and had some productive chats. Zoltan desperately wanted to be the Libertarian VP candidate—it would have raised his profile and his cause exponentially—and he felt like Johnson was taking him seriously. Not long before the Libertarian Party Convention, and the official selection of the Veep nominee, Zoltan was frantically texting Johnson, laying out the reasons he'd be great... "Be careful what you wish for," was the response he got back.

"He led me on a little bit with his texts," Zoltan says.

The transhumanist thought this was it—he was going to be chosen... He wasn't chosen. Johnson picked Bill Weld, another former two-term Republican governor—another one of the "old white guys," in Zoltan's words—to join him on the ticket. The two boomer career politicians pursued a more traditional strategy to win support across the 50 states, campaigning for more typical libertarian policies.

The strategy was a success, in relative terms. The Johnson/Weld ticket quadrupled Johnson's record-setting vote tally from 2012, and tripled the Libertarians' percentage share of the national vote.

But what might have been if Johnson had taken that radical path... The scorned interviewee was crestfallen. "I was devastated," Zoltan admits. Knowing he didn't get the gig, he drowned his sorrows in his favorite scotch.

(One thing Zoltan never mentions in our interviews, but which is a fact: While becoming the Libertarian VP nominee would be a massive win by itself, in the wild scenario where Johnson became president via a vote in Congress, Zoltan would still never have become VP of the US. In its vote to elect the vice president, the Senate can only consider the top two VP candidates in the Electoral College—in other words, only the Democrat and the Republican. A cabinet post or high-level advisor position for Zoltan was probably not out of the question, though.)

Zoltan still has fond memories of the evening spent with Gary. "We had a great time, just hanging out, talking. And the next morning we talked a bit more." He goes on: "I like Gary and I think he likes me." But Zoltan wasn't a libertarian in Johnson's mold, and that became clear pretty quickly. Zoltan was in favor of a universal basic income (UBI). To a libertarian who believes in self-reliance, trade, and exchange, the idea of handing out money for nothing was immoral. Johnson grilled Zoltan on his UBI views.

They shared a bond when it came to Ayn Rand. Johnson was intrigued that here was an Ayn Rand fan who could potentially draw in left-wing voters with his UBI idea, and his willingness to refocus American military and industrial power toward science and technology. At the same time, Johnson himself would appeal to more old-school libertarian and Republican voters. So, it could be a ticket that drew support from both sides and really changed the paradigm in the polarized US party system.

There were too many differences between the candidates, and too many other hurdles, for it to ever happen. It was all very well to say they could draw support from both sides, but they would first have to agree on a singular message and platform, one that could attract Libertarian donors and get approval from the party. Their campaigns' raisons d'être were wildly divergent. Johnson just wasn't invested in futurist topics the way Zoltan was. He cared about solid libertarian stuff like privacy, cutting red tape, opening up markets, lower taxes, and shrinking government. A lot of that didn't sit well with

Zoltan's appeals for drone surveillance, massive government investment in tech, and so on. Although the pair had a lot in common—their reading preferences, extreme sports (Johnson once climbed Everest), a history in the real estate/construction business, a fondness for marijuana—ultimately there was a divide in priorities they didn't bridge.

In an email to me, Johnson summed up the whole experience this way: "I like the guy but should have recognized before the meeting that his politics are not libertarian at all." He added that he "certainly disagreed" with Zoltan "on his contention he'd have brought more votes for the ticket."[9]

"I learned a lot from Gary and getting to know him and his team," Zoltan says. Disappointments come thick and fast in politics and you have to be able to roll with them and refocus your goals. "My campaign continued to grow anyways."

But to imagine—how might things have been different if a few choices had been made differently?

***

They filled out their mail-in ballots together, sitting in the office in his parents' Oregon beach home. Zoltan voted for himself, and his mom and dad voted for him too.

For his parents, the fact that they could cast a ballot for their son in a US presidential election... It didn't matter that he had no chance of winning... Given the life journey they'd had, this moment was immense.

The Science Candidate's campaign had officially ended the day before.

On the eve of the election, November 7, Richard Dawkins' Facebook page shared an article by Zoltan, and decried the lack of influence of atheists in presidential elections... Dawkins had boosted Zoltan's campaign several times on social media... but it was all to no avail. At least in terms of winning elections.

Not even celebrity endorsements like this could change his fortunes.

"As you get towards the last month or two, everybody realizes that all this enthusiasm spent towards third-party candidates is completely irrelevant," Zoltan states.

The Johnson/Weld ticket, officially third in the race, won no Electoral College votes, and got 3.28% of the national vote. Between them, the two major-party candidates—whom polls showed were the two most hated presidential candidates in US history—received 93.95% of all votes cast.

So it goes...

Zoltan is glad he got to finish the race as the Transhumanist Party candidate (by reputation, if not officially by electoral law) and as "America's Science Candidate," rather than someone's VP selection—even though he would have taken that job in a heartbeat. It meant he got to run an entire presidential campaign, for two years, under his own steam and on his own terms. That's become part of the "legend" he's built up around himself since then, something he capitalized on in the whirlwind his life became post-2016.

It's a cliché to say it, but no less true for it: We all face periods of death and rebirth in our lives. The next few years held both for Zoltan.

The day he voted with his parents was the last time he saw his father alive. Steven Gyurko died on May 5, 2017.

"He went from being a really, really fit man in his early forties," Zoltan reflects. From age 57 to the day he died, Steven had five heart attacks. After the fourth, the doctors said that he could no longer be operated on; his body wasn't healthy enough to survive. Officially, the cause of death was another heart attack. But: "My dad didn't die necessarily from a heart attack, his organs really just essentially shut down... We could see the slide, and there really was nothing to do... You have 40,000 kilometers or something of arteries in your body, and they were just all clogged... They wouldn't heal themselves." Zoltan says: "He died a natural death, that's the problem."

Zoltan talked to his father about cryonics, and offered to pay to have his body preserved, so that one day father and son could be reunited. But the elder Gyurko didn't want to keep on going or to return. His body had tried his patience enough. He was tired of struggling just to stay around.

"When you're physically not your best, or not even close to it," Steven had told Zoltan, in a son–father interview that aired online during the campaign, death becomes "like a solution to the problem."[10]

Mourning his father, Zoltan renewed his vow to avoid the same fate.

"My mom said my body is shaping up to be exactly like his... Unlike him, I really need to stay healthy and skinny as long as possible, because if I have some of these genetic factors, then I have a really tough fight."

Throughout our early interviews for this book, Zoltan had a persistent cough. But he resisted seeing a doctor because of what he might discover. That he was pre-diabetic. That he was prone to heart trouble, like his dad. That he would need medications. That his insurance premiums would skyrocket.

Wouldn't it be better to face those things now and act, rather than wait for a mortal shock? He promised he would visit a doctor soon.

The years after the presidential run pushed out time and space for anything but the transhumanist cause, and his family. He was in demand. His career took on a new life as he became a jet-setting speaker and commentator. The post-election world was very different for him, Zoltan says.

Within days of the end of his campaign, he was on a flight to speak at a global innovation summit in South Korea. He and his family were flown out, all expenses paid, and put up in a fancy hotel.

Zoltan's campaign had received good international coverage. US elections always make news around the world, and the transhumanist's run had novelty value. He was talking about issues no one else was, stuff that probably sounded batshit to the average reader or viewer, no matter where they were in the world, and made for good copy. Zoltan thinks he benefited internationally—both during and after the election—from the fact that the US electoral system was less understood outside America. The casual international observer didn't necessarily know that the Republican and the Democrat were the only viable candidates. "Zoltan Istvan, US Presidential Candidate" sounded a lot more illustrious to folks who didn't know that literally hundreds of lunatics with no chance threw their hat in the ring every cycle.

So, Zoltan's reputation internationally (at least among the media) was maybe higher than it was in the States. He'd got great coverage from Chile to the UK and Portugal, Japan, China, South Africa, and beyond... He received great coverage in his ancestral Hungary, too. Undoubtedly, this was the first time many readers and viewers had ever heard the term "transhumanism." He was introducing the philosophy to people in Indonesia, Albania...

Invites to speak at tech and innovation events came from all over the globe, and many of them were all expenses paid. He was hobnobbing with top businesspeople, political influencers, and politicians themselves. Getting a taste of a First Class Life. And truly getting a chance to impact policymakers. He spoke at a World Economic Forum event in Dubai attended by the prime minister of the United Arab Emirates and by WEF founder Klaus Schwab. Chatted with Chile's president at Congreso Futuro, Latin America's largest science event.

He was now a bona fide global activist. I joke that he's the very definition of the globalist elite. He's spent much of his post-2016 life traveling and speaking at events, and in every location he appears, he does media—and spams all his appearances across social—spreading the message of an amazing future beyond humanity, available to anyone, anywhere in the world.

Zoltan's 2016 campaign morphed into a perpetual campaign, a life devoted to advocating transhumanist issues through meeting people, speaking, writing, being interviewed, and bringing attention to those issues in whatever way he could.

His talk at Congreso Futuro 2018, organized by the Chilean Senate, was even attended by the executive director of the Nobel Foundation. Zoltan is giddy when he thinks about it.

Somebody is going to win a Nobel Prize for life extension, he says. "The person who can help conquer death is going to be responsible for saving billions of lives." The only question is whether the Nobel will go to the scientist who invents the method of longevity, or the activist who pushes society to make it happen.

Zoltan certainly sees himself in the running: "I feel like I stand a good chance of winning a Nobel Prize."

Chapter 11

Plague Politics

It was still night—he woke up—*my god, that pain!*

It was his chest. His heart? *Fuck!* He was sweating all over—all he felt was pain.

Zoltan tried to get out of bed, stand up—he fell back.

He managed to get up... made it to the couch in the hallway... sat down again.

He was clutching his chest.

He was coughing now—uncontrollably.

A million thoughts were going through his head.

His dad had his first heart attack in his fifties. Four more and then he died. Zoltan was 46. His mom always said he was developing the same body type and temperament as his dad.

Was his number up?

He didn't want to see a doctor. Hadn't been to a doctor in 20 years. He knew it wasn't smart. But what if he went to a doctor and they said he was pre-diabetic, had heart trouble, and he needed to go on a load of pills? His insurance would skyrocket. His life would get more complicated. He had just kept going.

Besides, his wife was a doctor. She would know if something was seriously wrong with him, right?

He hadn't woken Lisa. *Thank god.* He didn't want her to panic.

He should probably go to an emergency room. *Jesus.*

No—it's OK. I think it's getting better, he told himself.

He was just sort of going through it now, lying on the couch... It still hurt like hell.

The pain was easing, though. Maybe.

Fuck!

***

Zoltan barely slept that night and woke up the next morning feeling awful. His head pounded; he was clammy. He thought he might have had a stroke.

It was October 2019. In two days, he was due to go on an extended trip to Europe with Lisa and the kids. It was this fact—wanting to be right for the trip and the speech he was giving in Portugal—that finally sent him to the emergency room.

They did all the tests—EKG, blood pressure, treadmill, blood work, troponin test—and there was nothing wrong with him. He was maybe 15 pounds overweight. But he wasn't pre-diabetic. His heart was good. He was very healthy for a man his age.

Probably stress had caused that night-panic, they said. Sometimes you go on full-steam-ahead for so long, your body just tells you it needs a break.

He might have had an embolism, just a shock to the system that quickly righted itself. There was no sign of any blood clots or anything sinister now.

People die from this stuff, Zoltan thought. You hear about it all the time: A strong, healthy college football player drops dead on the field from a stroke. Sudden, inexplicable deaths. He'd been working so hard to stay alive, it could have killed him.

Well, maybe. Maybe not. Perhaps it was just random—nothing to do with his workload or stress or the increasing amount of scotch he'd been drinking every night.

That was the scary thing. Could he control it? Could he not? Predictable, engineered robot bodies were needed now more than ever.

He still felt like shit; physically weak and in a brain fog, constantly. That should all pass soon, he was told.

He went to Europe. He and Lisa were looking at vineyards in France to add to their property portfolio. They already owned vineyards/wineries in Argentina and Napa Valley. Zoltan wanted to grow the first wine infused with nootropics, "brain fuel." Wine with a transhumanist theme. He would probably call it The Transhumanist Wager. It was a business idea he was exploring as he thought about the next field he could expand transhumanism into... It could get him coverage in food and wine magazines around the globe, get the word "transhumanism" and his ideas in front of a whole new set of people... Goodness knows it was becoming harder and harder to receive coverage in the news media in the era of Trump, with impeachment talk, trade wars, and America's divided union crowding out space for anything else.

He gave the speech in Portugal, at the Business Transformation Summit in Lisbon. But it took all his strength to stand up straight on that stage. Throughout the week they'd been in Europe, he'd had four more "attacks"—moments of hyperventilating, extreme dizziness, not able to stand or see. He wasn't sleeping. He had to spend hours in bed during the day. He continually felt groggy and low-energy.

He actually apologized during the speech—because he wasn't keeping his train of thought and didn't have his usual verve. He lied to the audience and claimed he had the flu.

On stage, Zoltan was using most of his energy just trying not to collapse.

He was getting paid a lot of money for the speech. He also had to appear on a panel afterward. There was a large crowd and a ton of press in attendance. There were "a lot of people counting on me," he says.

He felt so bad after his talk that he forced himself to the emergency room again—this time in a foreign country. They did all the same tests in Lisbon as had been done in California.

Same result. His heart looked good—it wasn't a heart attack. Probably stress brought on by overwork.

Zoltan tells me: "What I think now happened is that I had some kind of lung infection that was coupled with maybe a brain embolism of some sort."

I remembered that he had been coughing throughout our interviews...

The biggest casualty was his confidence. He was used to his mind being sharp, to making a lot of quick decisions every day; Zoltan now slipped in and out of a brain fog for weeks. That hit his creativity and his productivity hard. Almost every day, on and off, he found himself physically shaking. *Would he ever get back 100% of the mental and creative capacity he'd had just a few short weeks ago?*

He sent me a private message right in the middle of all this, along with a photo of himself attached to tubes and wires in the ER.

"I don't need to say this to you, but life is very fragile," he wrote.[1]

But... All the symptoms, including the brain fog, did pass. All it took was a little more time.

When I spoke to Zoltan on Skype a couple of weeks after his return from Europe, he was still shaken by the experience, but ready to get back to work in earnest.

After a run for California governor as a Libertarian Party candidate in 2018—which had left him scarred—he was running for president again. This time, as a Republican.

Post-2016, Zoltan had decided he should infiltrate an existing political party and try to piggyback on its infrastructure for future campaigns, rather than run as an independent again.

(He summed up his 2018 experience to me like this: He had thought his association with Gary Johnson would endear him to the Libertarian Party of California, and he would have an easy time getting the endorsement for governor. But it didn't turn out that way at all. It started out pretty well. He appeared on editor Nick Gillespie's podcast on major libertarian site, *Reason*. Gillespie even later endorsed Zoltan's run on Twitter. But that was one of few high points. A raft of articles and blog posts and podcasts were published by libertarians questioning Zoltan's libertarian credentials, since, among other things, he wasn't opposed to all taxation and appeared to be in favor of driver's licenses. What's more, he didn't think the government stopping you from owning a tank was necessarily a sign of tyranny. With the media focused on Zoltan's transhumanist views, and libertarianism a side show, a lot of Libertarians decided he wasn't good for the party. The prominent *American Conservative* website ran an article with the headline, "Transhumanism Is Not Libertarian, It's an Abomination." In the end, he received the gubernatorial endorsement by just ten votes, 45 to 35, at the Libertarian convention. Another candidate, Nickolas Wildstar—whom Zoltan had disparaged in the *Daily Caller* as "an unknown rapper with a tiny social media presence"[2]— received the party's unanimous endorsement. For the first time, California Libertarians had endorsed two different candidates for governor. Zoltan was angry that party members seemed to be far more focused on ideological purity rather than winning elections... But, in fairness to the Libertarians, it wasn't at all clear that Zoltan was the electoral juggernaut they should be looking to, given his grand total of 95 confirmed votes as a write-in candidate in the 2016 presidential contest.)

Zoltan protests that he had been told by Libertarians that the party had "all this infrastructure" to support his 2018 candidacy. But he didn't find this to be true, at all. "It's very loose knit, there's very little money," he says; "it's just a bunch of cowboys fighting for freedom. And I like the idea that they fight for freedom, but these are people that can barely run conferences."

He decided he needed to think bigger when it came to his next campaign. The Libertarians had hated him, claiming he wasn't a true libertarian... If he was going to be hated in whatever party he joined, he was better off in a party that represented half the country, rather than a small percentage of it.

"Why be hated by such a small party?" he asks. "If I'm going to be hated, I may as well be hated by the Republicans."

Now, he was set to run in the 2020 primaries against Trump, as a new kind of Republican focused on "Upgrading America."

Throughout our Skype call, he seemed to grow in confidence as he went on describing his upcoming campaign.

He might be finally escaping the psychological trap of thinking he's doomed by genetics to his father's fate. Maybe he tends more towards his mother, he reflects.

I ask him how Ilona is doing. Less than two months before, she had moved in with Zoltan, Lisa, and the kids. The house she'd shared with Steven in Oregon finally sold.

Zoltan was looking after his mom (financially), but Ilona was helping out with the kids to give Lisa and Zoltan a break too. It was a win-win.

Zoltan told me he actually had to jump off our call to pick up the kids soon... Normally these days, his mom does it. But she hasn't been feeling well of late, either. It's just the flu, most likely.

There's a terrible virus going round the family these past few days, he said.

***

"I caved in and sold Ayn Rand to the devil."[3]

I got this message in the middle of the 2020 campaign.

When Zoltan and I first spoke about his policies for his 2020 run for president, he was bombastic. He was rediscovering the spirit of Howard Roark—someone who was entirely himself; letting it all out there, whatever the controversy. He was channeling Jethro Knights, and Jethro's belief that only the functional should be rewarded and survive.

Zoltan was focused not just on big ideas but on specific policies, targeted to attract Republican-leaning intellectuals and press. He'd hired a campaign manager to help him out, Pratik Chougule, a former executive editor at *The American*

Conservative and veteran of Mike Huckabee's and Donald Trump's 2016 campaigns (where he'd become disillusioned with both). Their thinking was that if Zoltan created enough of a buzz, he wouldn't, of course, beat Trump to the 2020 Republican nomination, but he might collect enough allies that he could later run for Congress as a Republican, or run for some other smaller office. Zoltan was ambivalent about holding any political office that wasn't on the level of president or governor. Dealing with people's small-ball neighborhood concerns would be completely soul-sucking. But he figured the strategy was worth a shot.

A book of Zoltan's essays was going to be central to the publicity strategy this time around. He was busy trawling through the literally hundreds of articles he'd written over the years, for dozens of media outlets. He planned to collect a range of his articles into a series of self-published nonfiction books on different political and philosophical themes. It was a way to set out his stall: Lay out all his ideas, thought-experiments, and potential policies in one easy-to-find place for those interested in transhumanism and potential voters. He published the first collection, *The Futuresist Cure: Notes from the Front Lines of Transhumanism*, in summer 2019. It was an introduction to the "what and the why" of transhumanism — what it entails and why you should care. He released the second collection in October, *Upgrading America*. This was basically Zoltan's campaign book. It included his essays on why we shouldn't fear automation; how leasing out federal land could pay for a universal basic income; how artificial wombs could bridge the abortion divide; replacing prisons with drone surveillance, and more.

Universal basic income… pro-choice… ending incarceration… These don't sound like Republican policies.

He was running as a Republican because Republicans needed the biggest wake-up call about the world of the future, Zoltan explained. He was also adamant that he was a fiscal conservative and economic liberal—in favor of low taxes (no new taxes in the case of his UBI proposal) and in favor of light-touch regulations on business; two things traditionally associated with the GOP.

This 2020 run by Zoltan was much more overtly about political calculations, as opposed to his 2016 presidential run, where "candidate for president" was really just a publicity stunt. This time, he didn't want to create a circus. He wanted to be taken seriously.

But, in order to stand out, he knew he still had to be a provocateur.

His first thought about how to do that was to return to familiar ground: He would channel Jethro Knights.

In a chat with me in September 2019, Zoltan summed up his and Chougule's thinking: "We feel like, in order to make a mark, I shouldn't try to be more diplomatic."

At the time, Zoltan was agonizing over whether to include several unpublished essays in *Upgrading America*, the book his campaign was going to "lean on" for policy heft. These essays were more "extreme," Zoltan told me, chuckling. They were more Jethro than Zoltan. They included a proposal that only those who make enough money to pay income tax would be eligible to vote; a proposal to essentially make homelessness illegal; and an essay about how the #MeToo movement had gone too far, because good men are now afraid to be alone with a woman.

"The Republicans will love this," Zoltan told me, referencing the income tax/voting idea in particular, "because it probably would benefit them." For him, it wasn't about who it benefited electorally; he did really believe on some level that if you weren't producing enough to contribute to the collective pot in America, you shouldn't get a say in how the system was run.

As someone who had built up his own wealth and contributed a lot to the collective pot over the years, Zoltan viewed the essay as his personal fightback against the rhetoric of Bernie Sanders and others on the left who constantly blamed the rich for the shape of the world.

In the end, all those previously unpublished essays made it into Zoltan's campaign book. The essay collection was a campaign cry to preserve America from the socialists and allow capitalism to be unleashed to bring about the transhumanist future... He didn't want America to become a social democracy, like the European model, where there was lower economic growth but higher social spending. Europe had benefited from the US's willingness to power the world economy—benefited from the inventions and innovations of modern life conceived in America—and benefited from the US's willingness to shoulder the burden of defending the West... Without America, Europe would not have the wealth or freedom to be so generous to its citizens.

Two months later, after his health scare, Zoltan still had not officially launched his campaign, but he had changed his tune.

When we spoke in November 2019, Zoltan reflected with horror on some of the political positions he'd been taking in September 2019.

What had happened with his body, his health crisis, he said, had made him "a little bit of a nicer person." He still wanted to be true to his values, which he described as "libertarian paternalism... this idea that libertarianism needs to be forced down people's throats." But he didn't have to be "an asshole" who stops poor people from voting. His experience over the last couple of months had reminded him of something important. That maybe it was better to be Zoltan than Jethro.

"You know what, you have to always try to remember the other side of the coin, which is, when people are down, they need help... You're not just creating policies for when we're all super-healthy and doing great...

"You can never forget this—you can't forget that there are moments that you are totally weak yourself, and you want to create policies that cover the entire spectrum."

He now wants to be more like Obama than Trump, he says. Someone who tries to unify rather than speak to one section of the country. Though he's still running in the party of Trump.

But... Even though he left that essay about taxes and voting in *Upgrading America*, he didn't make "can't pay, no vote" one of his official campaign policies. Instead, he proposed abolishing income tax entirely, and replacing it with a national sales tax, which everyone would pay when they bought stuff, whether it was a trinket at Walmart or a luxury car. (This was an idea that had been around in Republican circles for a long time. Indeed, it was again introduced in Congress after a new Republican majority was elected in 2022, as part of the Fair Tax Act, which of course didn't pass.) Zoltan still believed in the principle that you should contribute to the federal system you benefit from. But this way, everyone would contribute, and he wasn't pitting everyone else in the country against the 47% who didn't earn enough to pay federal income tax.

This wasn't the only topic on which Zoltan was having a "Road to Damascus" moment. He was also changing his tune on the European social system vs. America's. He was beginning to sound like Bernie when he compared the US and European medical systems, based on his recent experience.

"When I went to the ER in Portugal—this is crazy, man, you guys in Ireland probably have the same situation—the EKG and the run-up and the tests cost me €238. The exact same treatment in America cost nearly $3000. So, more than ten times the amount.

"And the service in Portugal was virtually identical—I went to one of the best private hospitals in Lisbon."

He laughs. "I'm not sure why it's ten times more expensive in America. I believe in capitalism, but I gotta say, something is wrong with ten times. I could understand three times!"

He hadn't yet worked out his policy for dealing with that.

"I don't want to go down a socialist or a liberal path in terms of telling doctors and hospitals how to offer care, but something is wrong when it's ten times more expensive and it's the exact same tests," he repeated.

He's thinking, no doubt, about how his parents lost a huge chunk of their net worth to his dad's medical bills...

"The amount of bankruptcies that happen because of medical bills in America... Is it really useful for people to lose their homes when they have a medical problem?"

It goes too far when people lose their houses and end up on the street due to medical bills, he says. Losing your house and downsizing to an apartment might be acceptable. There should be a basic minimum below which you can't fall. A social safety net. This is also why he campaigns for a UBI, he reaffirms.

All that said, there were still plenty of elements of "libertarian paternalism" in his campaign message, for sure.

He was running a much shorter campaign this time than in 2016. He'd be launching in late 2019 and aiming to make some noise through several Republican primaries, ending his run after a few months. A short, focused campaign would keep his costs low. A short campaign was also the only kind Lisa would agree to.

He needed eye-catching policies to have any sort of an impact. During that chat in November, he talked me through the "Zoltan 2020—Upgrading America" platform he was about to launch.

"We're going full out—we're going bold... We're going to licensing children... We're going to, literally, gutting the Constitution... We would offer large tax incentives, or financial incentives one way or the other, to people who remain within a certain type of weight—and eat a certain type of food and retain a certain kind of fitness level."

He adds: "We're going to pay you to be healthy. And if you don't want to be paid [extra], because you want to eat McDonald's hamburgers all day long, that's OK."

There's more: "We're definitely going to rewrite the disability laws... So you can't remain on [government] disability [payments] if there's a way in the future—let's say five, ten years—to fix your disability."

This was all in the service of promoting longer and healthier and fuller and more able lives... Radical, discriminatory incentives that pushed people in that direction.

If folks figured fuller and longer lives were good, maybe they'd start demanding upgraded bodies and eventual immortality...

As he wrestled with launching his second presidential campaign—the amount of personal time and money he would put into it; the struggle of trying to get on state ballots in the Republican primaries—Zoltan was also trying to find the impossible political sweet spot between doing good, staying true to himself, and being popular.

It was when the campaign got going, and he saw an opening to become more popular, that he ditched—overnight—even more of the contrarian policies, and he sent me that message: "I caved in and sold Ayn Rand to the devil."

Libertarianism was out, individualism was out; socialism and pleasing the masses were in.

You see, it turned out, the biggest buzz around Zoltan's campaign did not come from the media, even media outlets attacking him—as it had in the past. The media this time around were too fixated on President Trump's impeachment trial and the hectic Democratic horserace.

No... The largest share of attention Zoltan got in 2020 came from online fans of another candidate—a candidate in the Democratic race. These fans saw Zoltan as standing for something similar on the Republican side.

Almost as soon as he'd made the announcement that he was running for the Republican nomination, in November 2019, Zoltan received a question from a commenter on Facebook: "How do you differ from Andrew Yang?"[4]

Zoltan got a lot of love and interest on social media from #YangGang, devotees of Mr A. Yang, the entrepreneur generating a huge buzz in the Democratic primary by talking about how tech was going to eliminate jobs and how we needed a universal basic income to compensate. Zoltan was the Republican Yin to... well, you get where I'm going.

Zoltan's first approach to this attention from Yang fans was to suggest that he was the original and best Yang. He suggested what Yang had done was "borrow" Zoltan's 2016 campaign policies.[5]

Trouble is... When #YangGang looked into Zoltan's policies, they discovered UBI was about all he had in common with Andrew. And though interest in another candidate supporting UBI had brought #YangGang to Zoltan's door, they weren't going to stay for the rest of his policies. Heck, he wanted to license parenthood and stop welfare payments to disabled people who wouldn't "fix themselves." Fascist stuff.

Soaking in all this attention, Zoltan couldn't help himself. He wondered how much farther he could take his run if he played to the crowd.

He decided to go pure populist. He would scrap his principles to seize his chance for more popularity. (After all, the most important principle, the philosophy of the omnipotender, was to gain power.)

Zoltan declared on Facebook and Twitter, just a couple weeks before Super Tuesday (March 3) 2020, that he had a "major announcement":

Because my campaign is growing dramatically, I have—after much debate—changed many of the policies in my plan for US president. A politician aims to serve the people, and after listening for months to my supporters (and objectors), I have rewritten many of my ideas to more closely align with what the public sees fit, objective, and also just.[6]

The "public" he was referring to was mainly Democrat Andrew Yang supporters on Twitter. They didn't like Zoltan's anti-equity policies and views. So, since this was now his major support base, he just scrapped those policies.

You could see it as either refreshing or hilarious. Zoltan had a habit of "saying the quiet part loud" when it came to the naked ambition of his lust for power. Most politicians will at least pretend that they stand by their principles and don't bend with every wind-change in public opinion. Zoltan didn't pretend any such thing. He openly admitted he would make any concession he could to potential supporters if it meant he had a better chance of winning an election.

Zoltan had said the same thing in a *Newsweek* article a couple of years before. Of his 2016 presidential run, he wrote: "With leadership comes some compromise, and I veered both right and left (mostly left) to try to satisfy as many people as I could."[7] He was doing the exact same in 2020.

He told me explicitly: "The ultimate principle of mine is not to uphold a certain set of standards, the ultimate principle is not to die, so if I need to get into a position of power by lying my way through it—unlike John Galt—I will do so."

When Yang supporters couldn't understand why, or were angry that, a supposed advocate of good things like a UBI would run in the evil party of Trump, Zoltan reiterated that, essentially, he was running as a Republican because it was the worst party, the one most in need of the change he was advocating. Like a missionary, Zoltan would bring his message not to the faithful but to the most benighted: "The fact is the GOP, more than any other party, needs to open its mind about the future and also hear about transhumanism... I am slowly getting conservatives to hear about the future." He added, describing himself: "Some thinkers and activists are beyond partisan politics."[8]

Yang dropped out of the presidential race on February 11, 2020, following poor results in the Iowa caucuses and New Hampshire primary. Zoltan made a special appeal for funds that day on social media: "Every dollar helps me tell the Republican Party we need #UBI."[9]

A day later, he posted that he and his team had "tens of thousands" of extra folks interested in his campaign in the previous 24 hours, as Yang followers went in search of any candidate still in the race who supported UBI.[10]

Less than a week before this, Zoltan had defiantly defended his campaign's stance that becoming a parent should require a license. He shared a *New York Times* article about a father who was investigated four separate times for child abuse, before his baby was found dead. Zoltan wrote sardonically: "It's crazy that people suggest we consider licensing parents so they can have kids... Let's defend adults and their right to have as many kids as they want so they can make millions of children suffer unbearably each day... #MyPresidentSupportsLicensingParents."[11]

That was on February 7. On February 15, licensing parenthood was out of his campaign platform. Along with most else that had a whiff of controversy. Instead of requiring parents to get licenses, Zoltan's government would now provide training programs, "to try to get all Americans to be better parents."[12] Zoltan wrote that he hoped everyone would "discover more electability" in his rewritten 20-point plan.[13]

In the course of writing this chapter, I got into a debate over the phone with Zoltan's campaign manager, Pratik, about whether Zoltan's fondness for U-turns is a political asset or liability. Pratik didn't see it as a U-turn; Zoltan was just evolving to his electorate. He saw it as a mark of Zoltan's open-mindedness. "I think you're right, obviously," Pratik told me, "that aspiring politicians by nature realize they have to change course and so forth." But a lot of people, especially in fringe movements, he said, "go in the other direction, where they become very, very dogmatic and ideological and stubborn. And from my perspective at least, I always found Zoltan to be a very non-dogmatic, contrarian, and very open-minded political thinker." This was a plus for Pratik.[14]

"I can't in any way say that we did poorly," Zoltan says of his third failed election bid. "It didn't succeed in any way in terms of getting Republicans to take me seriously... Where most of my popularity came from was still with the left."

But... Zoltan says he doubled his Google search traffic on Super Tuesday 2020, versus Election Day 2016. Most of this was down to the fact that he managed to get onto the ballot in a few important state primaries, including California and Texas. Curious voters would have seen his name on their ballot paper and looked him up. (In 2016, his name did not appear on any ballot paper — you would have had to have heard he was running in advance, in order to know to write in his name on your ballot.)

And he did get some substantial mainstream media coverage in 2020, even if this aspect was dwarfed by his 2016 run. Journalist Jasper Hamill reported on Zoltan's run "against Trump" in the UK *Metro* — Hamill had covered Zoltan for years for different British papers. Jennifer Harper — another who'd covered Zoltan for years — did several articles for the *Washington Times*, including one where Zoltan was hailed in the headline as a "new type of Republican" (Zoltan's own line about himself). Thanks to *IEEE Spectrum Magazine* executive editor Glenn Zorpette, Zoltan did some writing for the Institute of Electrical and Electronics Engineers during the campaign. Over the years since he had become a force on the futurist scene, Zoltan had built up and relied on a war chest of media contacts like these to get his message out. Other journalists who he says opened many doors for him over the years include Gian M. Volpicelli, who was his editor at *Wired UK*, and Jason Koebler, editor-in-chief at *Vice Motherboard*. He also namechecks Tracey Follows, who wrote the definitive review of *Immortality or Bust*, for *Forbes*; Peter Clarke, author of *The Singularity Survival Guide*

(2019), who covered Zoltan many times; and Ford Fischer, co-founder of video news site *News2Share*, which has extensively reported on transhumanism.

Zoltan had tried to be provocative, he had tried to be popular... He'd stuck by his principles and sold out his principles. Neither option had gotten him very far in politics. He wasn't sure what lesson to take. Maybe there was none.

Pratik did have a lesson for Zoltan, but it wasn't one the candidate necessarily wanted to hear. Politicians are, according to Pratik, "effectively glorified fundraisers." Their life is the life of "a dial-by-phone salesman. They're just constantly asking people for money, begging people to host fundraisers for them, currying favor with wealthy people, and figuring out how to get donations." The candidates themselves have to do this—it can't just be left to staff—since the candidates are the ones donors are choosing to back or not back. Although Zoltan traveled in very affluent circles, he was consistently reluctant to beg for money for his political career. Until Zoltan got serious about this aspect of politics, he wouldn't make it to the big leagues, in Pratik's view.[15]

The 2020 run was good for his reputation and continuing to build his profile and expand his potential support base, Zoltan says. It was important simply to keep campaigning, keep running for things—so that if the stars ever aligned and transhumanism suddenly had a breakout moment, he was in the game.

The Republicans didn't like him. Democrats seemed to like him better. He had turned one way; he could turn back the other way, if he thought it would be to his advantage. So, he'd probably run as a Democrat next time out, see where that took him.

But of course... That's when something happened that made him return to his Ayn Rand roots once more.

***

I wonder if it'll take a Supervirus that kills tens of thousands to get the government to decide to spend its resources to eliminate major plagues.

Myself and other transhumanists are calling on politicians to realize that spending trillions of dollars on far-off wars, instead of spending a fraction of that money to potentially save our country from dire existential risks is plain irresponsible.[16]

Zoltan wrote that in a *Huffington Post* article in 2016.

In January 2020, I listened to reports of cities locked down in China, people told to stay inside their homes, after a new, super-infectious disease had been discovered. It seemed like a story from another planet; nothing that should concern us.

I remember taking a car ride with a colleague who'd just moved back to Ireland after a long time in Taiwan. He'd arrived in Taiwan shortly after the SARS epidemic had hit. He was scared. He knew the first wave was about to hit us. I didn't believe it—it just seemed so outlandish, so disruptive. Governments rushing to build emergency hospitals… People ordered to stay home or wear protective clothing—a mask or a visor—if they went anywhere… People dying alone because no one could visit due to risk of transmission… These were things that happened in movies—or, at least, on the other side of the world, far from my comfortable and routine life.

I remember sitting in work meetings in early March, getting the news that we might have to cancel a conference planned for April in the US as a travel ban was looking imminent: Passengers from Europe would not be allowed to enter America. My mouth dropped open—*a travel ban?* Then we got the news that trading on the New York Stock Exchange had been suspended due to a colossal market drop. It was surreal.

I Skyped Zoltan again during this time as the number of COVID-19 cases skyrocketed across the West.

He was in a weird sort of giddy mood. He was feeling vindicated. He'd repeatedly warned, in columns and on the campaign trail, about the existential threat to the human race from an event like a breakout virus. He saw this as a real chance to prove the case for transhumanism. Wasn't it obvious, now, that we needed to work harder to eliminate all disease and death—work harder to master nature, or it would be the master of us?

Zoltan was also, it seemed to me, getting a strange kind of thrill—or maybe it was just a survival-instinct adrenaline rush—from living through the real-life possibility of social collapse. Almost like he was getting to live through the climax of his own novel, *The Transhumanist Wager*, or Rand's *Atlas Shrugged*.

He was "looking at this from a very 'prepper' point of view," he told me. "I've always taken prepping very seriously because that's what the sailboat is."

When you're leaving to spend weeks at sea, you have to stock up and be prepared in order to survive. He drew on that muscle-memory now as the pandemic took hold. Just in case society ground to a total halt due to government-ordered lockdowns or people unable to work because they were infected. And the resulting social unrest.

"We bought generators, we got gasoline, big giant water jugs, I got ammunition and guns," he excitedly told me.

He posted on Facebook: "Friends, do not believe Government will be there for you or supply chains will remain intact"— along with a picture of a shopping trolley he had packed to the brim with canned and dried foods, cleaning products, and toilet paper. "Cherish your independence."[17]

He was ready for anything. He joked on our Skype: "Let the zombies attack!"

Zoltan's first-instinct, adrenaline-rush focus on preparing for the worst, however, soon gave way to confronting what was, in his view, the worst-case scenario.

What if everything that was being spawned by COVID... Government-imposed lockdowns, businesses forced to shut, millions out of work... What if the economic devastation meant that capital dried up for anti-aging and transhumanist-tech research?

What if... the engines of progress just stopped?

What if... Zoltan wasn't going to live forever after all?

He was so concerned, he rushed out an open letter to all "Fellow Humans," which he self-published on *Medium*:

> If you believe in the life extension movement of trying to live indefinitely through science and technology, then you likely should not support the worldwide quarantine... It's horrible that so many lives will be lost by COVID-19, but in a "worse-case scenario" it's likely 100 million people (a more likely case is about a million people despite what sensational media tells you) will die globally (mostly older people who have only a few years left to live due to their underlying medical conditions of aging—and who have likely been kept alive due to science and 21st Century medicine anyway). But the damage we could cause (and almost certainly are causing) with the quarantine and shut down to the US and global economy may cost the life extension movement and its scientific research possibly three to five years of progress— because the funding, projects, and jobs around the anti-aging industry will disappear for a notable time. The math shows that if we achieve indefinite lifespans for the human race by the year 2035 vs 2040, approximately 250 million lives will be spared and could then go on indefinitely.[18]

Zoltan and his family managed to survive the pandemic pretty much unscathed. They were affected by the same things everyone was: business closures, entertainment and hospitality shutdowns, school closures. Lisa took a pay cut at Planned

Parenthood. But they had a nice garden and grew their own food, and had the money to stock up early on things Zoltan thought would be valuable or they might run out of. In fact, ultimately, Zoltan used the pandemic to double his and Lisa's net worth—moving stocks around in the crazy market volatility, and purchasing real estate that would rise again as property prices took off.

And yet... He was whacked by a wall of pessimism. Zoltan's posts from the heart of 2020 read like a man whose world was imploding.

In other contexts, the struggle between Zoltan's accommodating liberal instincts and Jethro's blunt, function-over-everything mentality had already played out in 2020. When the pandemic hit, Zoltan's "Jethro" personality won out. Zoltan went for full-throated libertarianism.

He vocally opposed lockdowns. He openly advocated that continuing economic progress was a greater good than saving elderly people's lives—since those with "few life hours left anyway" were most at risk of death from COVID-19.

The thing is, Zoltan soon found that the stakes for his reputation and his career were suddenly higher now than they'd ever been.

He'd tried—through several election cycles—hopping from one side of the political fence to the other. The stakes involved in doing that had always seemed low, given he was unlikely to win any office. Media would cover his controversial opinions and nobody really cared too much, because he was writing about abstract, out-there, far-off-future stuff...

But the pandemic was real shit. People were dying, right now. Every single day. Hospitals were overwhelmed. Folks were trapped in their homes by government order. Nerves were on a knife-edge.

Zoltan had tried to tell his supporters, earlier that year, that his wide range of views placed him beyond partisan politics.

But, especially right now, America didn't want to hear that crap. You supported government restrictions to stem the spread of cases, and a rapid vaccine rollout... and that put you in the liberal camp. Or you were skeptical of vaccines and thought government had no business telling individuals what to do and where to go; that lockdowns and curbs on freedom were a "cure" worse than the disease. That put you in the Trump Country camp.

Zoltan was shocked at how quickly just airing opinions could potentially get him "canceled."

He was booed at a debate when he said that getting vaccinated was important.

Angering the "other side," he wrote a piece for the *New York Times* (he'd been writing for them occasionally for the past year) arguing that surfing should not be banned during the pandemic:

> I usually go to a popular reef break called the Patch, which is a 30-minute drive from my home here in the Bay Area... located just off the main beach in a reclusive hippie town called Bolinas...
>
> Once the coronavirus came, so did a Bolinas lockdown. Concerned citizens and government officials put up signs along the road into town, reading "Surfers Stay Home, Save Lives" and "Beaches Closed Due to Covid-19." Some residents even stood along the road and yelled at cars with surfboards on their roofs to turn around.
>
> I don't think many surfers obeyed. I didn't. I just couldn't see how walking out of my house, getting into my car, parking near the beach, and paddling into waves could be dangerous for anyone.[19]

Almost as soon as the piece was published, Zoltan was attacked — and not just on social media. Locals in Bolinas printed up posters picturing Zoltan with devil horns and the words, "Go Home Zoltan #kook."

Zoltan wrote on Instagram that he "received numerous violent threats…from Bolinas locals, including a death threat," and "50+ comments…warning me never to come back."[20]

I sent him a private message joking that he was "the Most Wanted Surfer in America." He messaged back that the whole thing was "quite funny":

> This Bolinas place is all hard core Bernie Sanders supporters, all white, and all entitled little pricks. They gave me [a] lot of threats but many of them have never seen me, and don't realize I'm a convicted felon, 230 pounds, spent my time in war zones… It won't go well for them.[21]

(There is a surfing subtext here which has nothing to do with the coronavirus. A lot of local surfers always object to folks coming in and using "their" beaches. COVID was a great reason to keep outsiders away, when you didn't want them there anyway.)

Still, in the hot-hot political climate of 2020, Zoltan couldn't put a foot right. He was called "another self-loathing white man"[22] when he advocated reforms to encourage more women and minorities to run for office… When he posted that Trump was "f—ed" because of COVID and the economic cliff-dive,[23] he was accused of being "just another communist in disguise."[24]

He worried his stance on vaccines would get him canceled with Republicans, while his stance on government restrictions would get him canceled with Democrats or the media. They'd call "time's up" on this "kook."

It came closest to that when Zoltan aired what he considered to be one of his least controversial views. In the wake of George Floyd's murder by a white police officer, as protests took place all over the US, Zoltan posted: "At least 5,000 kids (mostly black) starved to death today, & will tomorrow & day after, etc. Racism is terrible, but the massively bigger issue is poverty. If you care about change and are angry, DEMAND a universal #BasicIncome #JusticeForGeorgeFloyd."[25]

He added, the next day: "The riots have more to do with the lockdown (& how it stupidly forced inequality & massive job losses) than racism."[26]

That's when "real journalists" unfollowed him.

"I went online and I posted that I think the Black Lives Matter—the riots in America—have more to do with economic inequality than with systematic racism. And a bunch of my journalist friends wrote back and unfriended me. Like, real journalists at big places. And they were really angry that I was saying that it's economic inequality causing riots and not systematic racism.

"Again, I'm not denying systematic racism. I'm just saying, look, I know exactly what's happening—you can go to a place like Oakland, California and these people are very, very poor, they don't even know where their next [meal] comes from in some cases. In America. Of course they're angry—they're going to riot at the first opportunity.

"But what was really scary was, these high-end journalists literally unfriended me."

And so, after trying to fight his corner for a few months… His corner, in his view, being both pro-science and pro-individual rights—i.e. pro-vaccine but anti-lockdown—a stance true to his libertarian and Randian roots… Anti-racist but not looking at things through the lens of race…

After fighting this corner in the first half of 2020, and seemingly only alienating everyone… Zoltan took perhaps the most unusual step of all.

He decided to shut up.

***

The pandemic showed how far humans have come in our ability to control nature. Vaccines and therapies to mitigate the disease and save lives were available within a year—testament to the efficacy of science and the wonders of biotech.

The impact of COVID-19, however, also demonstrated how far away we as a species stood from the transhuman future Zoltan and his ilk hoped to usher in.

Far from being on the brink of immortality, recreating time and space, raising the dead, and ending all suffering, the truth was that our lives, our economies, our societies could all still be brought to the brink by a microscopic parasite.

A large proportion of the population would not even allow something that could protect them from disease to be injected into their body—because of religious beliefs, fear of needles, conspiracy theories, or personal choice... So, what hope was there that folks would soon start replacing their biological limbs with mechanical ones, by choice, as Zoltan hoped? Or get microchipped by choice?

We are far from the transhuman future, not only in terms of the science. As human societies, we're a long way from accepting the philosophy that it's good to be more than human.

Zoltan was bringing a message to America that very few people in America wanted to hear. As 2020 wore on, this became clearer and clearer.

Zoltan was completely out of sync with America's priorities—and got stick from all sides when he tried to wade into contemporary events.

If media on either the left or the right decided that Zoltan's views were now simply too obnoxious to air, he wouldn't have a home on the other side. How far would that set back the cause of transhumanism?

All of a sudden, riding out this volatile time, when even major stars could fall due to offensive tweets—riding out the age of "cancel culture" without being canceled—became a top priority. Zoltan stopped writing and posting about politics for a while, and just shared transhumanism stories.

The truth is, during one of America's most tribal eras—a cold civil war between woke Democrats and Trumpist Republicans— Zoltan had no tribe. He borrowed views and ideas from both sides of the divide and from neither; his politics were focused on creating a far-off moment in the future, and he hadn't enough to say about the present moment to draw either side to *his* side. That meant he couldn't count on anyone for protection.

No tribe... No place that's home... It wasn't a comfortable position to be in. But it's who Zoltan was: He didn't fit any mold. And he'd never liked staying in any comfort zone, anyway.

In a sense, today, he was like a solo sailor again... relying on his wits and his own judgment... charting his own course into an unknown future.

Chapter 12

To Meet the Stars as Equals

We will get to live in outer space—as transhumanists hope—but there's a lot that has to happen between then and now.

The Transhumanist Wager provides one vision of how we get there. The journey from earthbound present to unbound future. Though Zoltan may be like his fictional protagonist, Jethro Knights, in some respects, and unlike him in others, Zoltan's belief that this is the future he'd like to see is unquestionable. It's the future he's fighting for.

Two visions of humanity vie for supremacy in the novel: the conservative forces that think of humans as natural beings, products of God or Nature; stewards of a divine plan or a natural order... the earth itself and our bodies as sacred manifestations of an unknown force—not ours to corrupt with selfish desires.

And then, on the other side: those who believe individual humans have every right to seek to become more than what they are. Those who do not revere Nature, but see it as something to be improved upon. Who do not recognize any preordained order, but instead view life as a thing to be constantly reinvented, however they would like it to be.

In the novel, all politics—the struggle for what the future will look like—ultimately comes down to this: anti-transhumanists vs. transhumanists. Those who want to preserve the status quo or some version of life as it has been, who think there are higher values than personal preference. And those who think humans have failed if we're not always pushing the boundaries of technology and life itself; who see no higher goal than empowering each individual to choose when and how and if they die.

In the novel, of course, the transhumanists win the war for tomorrow. Zoltan still hopes that second vision of humanity is the one we all choose.

We should want to be free of Nature, not in harmony with it. That's the only way we will get to the endpoint of humanity liberated from its nature, as imagined by Jethro and the transhumanists in *Wager*.

At the end of the novel, Jethro states how he sees the human species evolving, a public pronouncement to the masses living under his new world order:

In less than a century, most of you will not be biological anymore. You will already be part machine, part cyborg, part whatever the evolutionary course takes for us to become stronger and more adept...

...

The species transhumanists create and evolve into in the near future will have the power and intelligence of thousands, if not millions, of human beings combined. And in that new, noble world, we shall find plenty to keep our interests hearty. We shall finally begin to see how big the universe truly is and how it really operates...

People of Earth, these are my final words: The moment has arrived for a new philosophy, morality, culture, and intellectual vibrancy for the planet, prompted by the leadership of transhumanists and a worldwide futurization of values...You may say we are monsters because we will soon amputate our limbs and cut out our organs, and replace them with better synthetic, robotic, and cybernetic parts. Or because some of us already have computer chips in our heads, enhancing our lives and behaviors. Or because we adhere to an egotistical, unforgiving set of ethics that favors the individual and the rise of undemocratic technology. But those of you who survive long enough into the future will all act and think like that.[1]

At first after *Wager* was published, Zoltan imagined he would spend the next few years writing sequels. He envisioned a trilogy. But then his life as a commentator and public speaker took off, and crowded out any time for novel-writing. There was always another interview, another event to prepare for. He was writing dozens and dozens of short articles based at least in part on all his ideas in *Wager*, trying to game search engine results for transhumanism. He figured all this was, at least for now, more important than completing his trilogy. A lot of people could write transhumanist science fiction. In the next few years, even chatbots would start doing it. But maybe, just maybe, only Zoltan had the right set of skills and the right pedigree to become the first successful transhumanist politician; win power, or at least influence the culture in a big way, and make real change in the real world.

***

"My wife is like, 'You should stop trying to run for president and just write these novels. It would probably be a lot better on your health.'" Zoltan laughs as he relays Lisa's advice.

They could do it—just decide to live another life. They had the money.

Although Zoltan spent most of his waking moments advocating for crazy futurist ideas, he had remained conservative with his wealth. Almost all his money was in property. Perhaps surprisingly, even if he supported their goals, he didn't invest big-time in tech startups or crypto or the metaverse or whatever the current futurist thing was. He was happy to give shout-outs to those things on social media and in any articles he wrote.

But was Zoltan actually going to put a lot of his money behind moonshot notions like this? No. He would stick to old-school bricks and mortar, the asset that had made more people rich than anything else in history. That conservative policy had meant he could continue to live a very nice lifestyle, and devote his time and his voice (if not his cash) to futurism, and not worry about having made a bad bet on a startup. His property wealth meant there were many options for the next chapters in his life...

Lisa and Zoltan had recently been chateau-shopping in Europe. They could move to France, live on their Bordeaux country estate with its own vineyard. What an amazing experience for the kids! Zoltan could make his nootropics-infused wine and write novels. Idyllic.

He's totally tempted by that other life, he says. But the chance that he could win a major election... Or a Nobel Prize for popularizing a movement that ended death... Those possibilities are too real for him to give them up.

If he dropped out of the public eye, completely out of the political sphere, even for a few years, to write books, he couldn't pick up where he left off. The world moves on quickly. He's achieved a position as "one of the world's leading futurists" — maybe *the* indispensable voice of transhumanism — "because I push and push and push," he says. It requires "constant work" to maintain his position in the public eye, "seven days a week, 12 hours a day," answering emails, responding to inquiries, sharing articles on social, accepting almost every request to do an interview or write an article. If he wasn't committed to his cause, Zoltan would much prefer to write novels or "just go surfing all day long," he tells me.

There are all kinds of other lives he would love to live if time wasn't an issue. Lives that have nothing at all to do with transhumanism. He would rework and put out his unpublished first book, about his travel adventures. He'd write a screenplay based on the months he spent hunting for Spanish silver at the bottom of the Pacific: the tension between the young divers and the rich bosses—the camaraderie and rivalry; the dream of finding the treasure and going AWOL with it, disappearing off to live among the locals...

The sail trip still looms large in his imagination, in how he imagines his ideal life. As of this writing, two decades since Zoltan abandoned the circumnavigation, *The Way II*, his 30-foot sailboat (he replaced the original sloop toward the end of his time at sea), still sits in storage in Greece. He never got to take it through the canals of Europe, across the Atlantic, round South America, and back to LA—the way he'd always planned. Zoltan pays the storage fees for the boat every month, even though he'll probably never take it out on the water again.

"I keep the boat because it keeps the dream alive," he says. "The dream of sailing around the world is very much alive to me, and I absolutely plan to finish it someday. Just, life is busy..." He'd even want to complete the trip when he's a cyborg, he says.

"Maybe one of these days I'll still do it." Then he laughs, thinking again. "My dreams of continuing on the boat are probably just dreams at this point."

***

Will we really get to the kind of transhuman or post-human future Zoltan imagines? Moving our brains from one body to another, whether it's biological or mechanical? Uploading our minds to the internet, and existing in full virtual reality? Creating children without "flaws"—flaws like genetic diseases, inadequate immune systems, or mental illnesses? Living however long we want, by resetting our age or switching bodies... even living forever?

Evolving—by our own design—into unrecognizable new forms of sentience, like living stars or invisible gods?

"There's no question we're going to get there," Zoltan assures me.

If we humans imagine our species doing something at some point in the future, chances are, we eventually will be doing it for real. That's what the record of history suggests. Leonardo da Vinci envisioned a flying machine centuries before it became real. *Star Trek* anticipated the flip phone. Space travel was imagined in all kinds of fiction before the launch of *Vostok 1*. Test-tube babies were presaged by *Brave New World*.

There is no doubt that scientific and technological progress has already transformed our species and continues to do so. But transhumanism has its critics in the scientific community, who argue it's based on a flawed premise or simply fake science. That it's just a modern-day quest for the mythical elixir of life, clad in scientific methods. Doomed, as all such quests are, to failure — since it seeks the impossible.

I'm reminded, however, that in 1903, just two months before the Wright brothers' first flight, the *New York Times* suggested that mathematicians and mechanics would not be able to produce a flying machine for at least a million years, and in any case, humans should devote their efforts to something more useful. In 1920, the same newspaper — widely considered the best news source in the world — argued that space travel is impossible, and mocked rocket pioneer Robert Goddard for lacking basic high-school scientific knowledge. The point is: The great and the good wrongly predict the future of science and technology all the time. Everything can't be done until it's been done for the first time.

The work of transhumanism continues... To create a medicine or therapy that slows aging, to weed out genetic illnesses, to link our minds to computers (so we can live in virtual reality), embed tech in our bodies (like Zoltan's microchip, but better), replace organic components with superior synthetics (organs that never fail)... All the different ways that could help us live much longer and better and different kinds of lives. Only time will prove transhumanists or their critics right on the science.

Zoltan, of course, is not engaged in scientific discovery or tech development. Zoltan is a philosopher, a promoter of a particular way of looking at the world: transhumanism as a moral ideal; something that it's right to pursue. He's devoted himself to asking people to think very differently about what life can be and what it should be.

Often, that means asking people to turn away from everything they believe—and consider a radically new view of what's good about being human. Not that we are made in God's image or that we're stewards of the earth. But, to use the Nietzschean phrase, that we are a bridge and not an end.

Confronted with his vision, it's normal that "everyone fights back," Zoltan says. Society's bias is toward the status quo: no changes that could radically disrupt agreed-upon ways of life.

Gene-editing tech... Cyborg limbs... These things are rejected, initially, as pushing beyond the bounds of what life should look like. Maybe some things are banned. Moratoriums are enforced. But the technology pushes ahead in the ways that it can. And, sooner rather than later, more and more fundamental change becomes accepted.

When IVF became available and the first "test-tube babies" were born, in the 1970s and 1980s, that technology was hugely controversial, and compared—in the UK parliament and elsewhere—to the science-fiction horrors of *Frankenstein*. For many years, it was a common view that interfering in the natural process of producing humans—via IVF—was at best an ethical minefield and at worst a grotesque transgression. Today, IVF is a normal and totally accepted part of fertility treatment. Over half-a-million babies are born via IVF every year. Any debate about whether this is ethical or not has moved entirely to the lunatic fringe. It would be grossly offensive to compare a baby born via IVF to Frankenstein's monster.

Once enough couples who'd had trouble conceiving realized that science had found a way to help them, the moral conundrum simply went away. Because real people were seeing real benefits. And the anti-IVF morality — which, ultimately, would have denied people their own children — looked ridiculously abstract, not to mention cruel, next to happy, smiling families.

The debates over the ways in which it's ethical to change human life are contentious, of course. But that's mostly when the outcomes are moral abstracts, not related to you personally.

When I was growing up, there was a story involving my disease which made headlines around the globe. Coloradans Lisa and Jack Nash, parents of Molly, a 6-year-old with Fanconi anemia, were told their daughter would die without a stem cell transplant. The big problem: There were no matching donors to provide the stem cells. So, Molly's parents and modern science created a donor.

The Nashes decided to have another baby. Their embryos were screened and one without the genes for Fanconi anemia was selected; an embryo that would also be a match for Molly. Molly's mother was implanted with this embryo, which became Molly's brother, Adam. Adam's umbilical cord provided the stem cells that saved Molly.

For a lot of religious folks and cultural conservatives, commenting in newspaper columns, this was another real-life case of Frankenstein making a monster that would wreak havoc upon the earth. The Frankenstein analogy always seems to be rolled out when advances in genetic research occur. These commentators were outraged that the Nashes made embryos that would never be implanted. They were outraged that the Nashes had deliberately *chosen* to implant an embryo that was healthy (i.e. didn't have a genetic disease) and that could save their daughter's life, rather than leaving those things up to natural selection. According to them, it meant Adam's first identity would always be as a "donor" rather than a human being existing for his own sake. It was a slippery slope to screening every embryo to rid the world entirely of people with Fanconi anemia, Down's syndrome, deafness, genes that indicated cancer—whatever we decided was a "burden."

For the Nashes, in contrast—the people actually involved, rather than pondering the story from the outside—this was a chance they simply had to take: They got to expand their family, giving Molly a sibling, while saving their daughter from imminent death. They got to give both kids the best possible chance at healthy and long lives, rather than watch one child die, and roll the dice on whether their second would suffer the same life-limiting condition. (Both parents have to carry an FA gene for a child to be born with the disease; if both parents are carriers, there's a one in four chance that any of their kids will have it.)

The ethical issues in this case were hotly debated across the world. The doctors involved received threats. The subject so ignited the imagination. At the time, we even discussed it in my high-school politics class. An anti-abortion group was quoted in the *Irish Times*, saying this kind of screening belongs on a farm. A commentary piece in the Irish press expressed outrage at discarded embryos left on the "dung heap of history."

"If you were in my shoes, I think you'd react a whole lot differently and the people that would continue with 'I'd let my child die,' I tip my hat to them," Lisa Nash said, 17 years after the events that propelled her family to fame. As reported by ABC Denver, Molly Nash was by then a happy, healthy, and ambitious 20-something; she and her brother, Adam, even closer than typical siblings. Lisa had dismissive words for her critics: "Good for you that you could watch your child die and not do anything."[2]

If you're a parent who has a terminally ill daughter, and you also want another child; if you can save your daughter's life and give her a sibling... Surely that's all upside? Surely, of course, you do that? You save or make better your child's life if you can. That's the only moral imperative. When the issues relate to you personally, the right thing to do is often crystal clear.

Zoltan agrees. The hand-wringing from politicians and ethicists about transhumanist tech like genetic editing will continue. But, ultimately, the public will accept and even demand these things, because the advantages are just so obvious: "We're talking about eliminating cancer in our future children. Come on, people, this is an OK thing!"

Philosophers and commentators argue in the abstract that suffering is good, that we learn from it. But you don't want your child to have a life-threatening illness so they can survive and learn something. You want your child not to suffer.

It's really the same with death itself. People argue that death gives life meaning. That, because our time is limited, we are more likely to use it wisely. That we value life as precious because we know it ends.

But... Nobody serious argues it's a bad thing that global life-expectancy has doubled in the last century, from less than 40 to more than 70 years of age. That living just 40 years somehow makes life more precious than living 80 years, and we should therefore aspire to shorter lifespans. Nobody seriously argues against efforts to give folks a better quality of life into their seventies, eighties, and beyond—a long "healthspan." Why should it be any different if we lived to 300 or 500 or 10,000?

In the abstract, philosophers and commentators and preachers argue that we need the specter of death to enjoy life. That we must accept aging and decay as part of the turning of time that gives each generation a chance to make its mark.

But if there were a way to reverse the aging process, maybe again and again and again, so we could effectively become young and healthy individuals forever, does anyone really believe that most folks would not jump at this opportunity?

We fear death. Most of us do not want to die, unless life becomes so bad that death is considered a release. Wanting to not die is considered the mentally healthy thing. Wanting to look your best, that is, as young as you can, and to be as physically fit as you can, for as long as you can, is considered an ideal way to live.

We make up stories about an afterlife, or death being part of some grand plan—something we must accept—just so we can make our way through the world. It's a psychological crutch.

But of course: If never dying became an option... If living in a version of your physical twenties forever became an option... If never losing a family member or a friend became an option... most people would want options like that for themselves and for their children.

Some of the more out-there transhumanist ideas may take longer to catch on: choosing to go with robot arms and legs; mind uploading; living in space... But some pioneers will want to take those steps first, because some humans are always driven to do what's never been done. And then others will follow, if they see the first ones survive and thrive, and if they believe taking those same steps will enhance their own lives.

Not so long ago, we'd think it impossible to routinely take organs from one human being and surgically insert them into another human being, crossing the boundaries of life in that way...

A hundred years ago, the idea that so many of us would regularly journey from place to place in flying machines, or spend so much time with our faces glued to tiny computers we carry around in our pockets... these things would have seemed bizarre. But here we are.

***

As I worked to finish this book, in late 2021 and into 2022, the pessimism that descended on Zoltan at the height of 2020 had lifted.

He was back to imagining the best when it came to the future. He was back talking about politics on social media. Even taking a few digs at cancel culture. Zoltan seemed to have thrown his lot in with the right, even though he was already talking about running for president in 2024 as a Democrat. In one post, he argued that mainstream media make "big money" on "witch hunts, which creates addictive tabloid-like news."[3] When the world's most popular podcaster, Joe Rogan, came under fire for spreading COVID-19 misinformation—and conservatives leapt to his defense—Zoltan criticized the "Soviet style" attack on Rogan's free speech.[4]

The past couple of years had brought quite a few milestones. Daniel Sollinger's documentary on the 2016 bus tour, *Immortality or Bust*, was picked up by Amazon Prime Video and made available for instant streaming. It also saw a DVD and Blu-ray release. The film had debuted at the Raw Science Film Festival in LA in 2019, where it won the Breakout Award. Zoltan, his mom, wife, and kids all joined Sollinger at the red-carpet premiere. This was one of those precious moments where Zoltan really did feel like his efforts were being recognized and validated. It was Sollinger's film, of course, but Zoltan and his wild pilgrimage around America were the subject. With the streaming and DVD release, that journey and the reasons for it could now be replayed in homes all over the world.

In 2021, the first book devoted solely to Zoltan's work by another author was published: *At Any Cost: A Guide to The Transhumanist Wager and the Ideas of Zoltan Istvan*, Chris T. Armstrong's 500-page tome.

Between 2019 and 2021, Zoltan self-published no fewer than seven volumes collecting his articles, essays, interviews, and other writings, a box set he dubbed the Zoltan Istvan Futurist Collection. The first volume, *The Futuresist Cure: Notes from the Front Lines of Transhumanism*, had a foreword by futurist Jacque Fresco. Fresco, who died in 2017 aged 101, was famous for establishing the Venus Project with his partner, Roxanne Meadows—an idea for a utopian city where robots provide for humans and there's no money or private ownership. Zoltan had met Fresco and Meadows toward the end of his 2016 campaign, at their 21-acre campus in Venus, Florida, which showcases their plans. Fresco wrote in *The Futuresist Cure* that "Zoltan's work is helping to prepare" people for "a saner system for the future."[5]

In September 2021, Zoltan set off on a new academic adventure, starting a Master's in Ethics at Oxford University, studying under bioethicist Julian Savulescu, who controversially argued for engineering humans who are more likely to have good morals.

Oxford "is a sacred place and time in my life," Zoltan tells me.[6] Studying future ethics at one of the world's oldest universities, wandering the hallowed halls, enjoying candlelit dinners and the whole medieval university experience. He's getting to relive a youth he never had. And he gets to think and write deeply about how society's ethics need to evolve to accommodate the transhumanist future, with one of the world's foremost bioethicists as his thesis supervisor.

Another well-known bioethics philosopher, Peter Singer, had taught Savulescu—so Zoltan felt he was joining an august lineage. He was wandering the same halls, sitting in the same rooms, as Hobbes, Locke, Adam Smith, James Smithson, William Fulbright, a long list of thought-leaders and world leaders... Part of Zoltan's research would soon focus on the impact of ChatGPT and generative AI, powerful content-creation tools putting the jobs of journalists and artists at risk. (Zoltan himself finally decided not to write the sequels to *The Transhumanist Wager*, for now, after he asked ChatGPT to come up with some plot ideas, and the ideas the AI spat out were all better than his.)

The main reason Zoltan had stopped commenting on contemporary politics during 2020 was because he didn't want to put at risk the success of the documentary or the books—or, indeed, his entry into Oxford. (It was far from inconceivable. Jordan Peterson had recently had a planned fellowship at Cambridge University canceled because he was too controversial a figure.) But with a short break in the US's political fever in the early days of the Biden presidency—Zoltan had voted for Biden, by the way—Zoltan decided to try carving out a political niche for himself once again.

He didn't really see an alternative but for him to keep pounding the drum for transhumanism within the political sphere. Policies needed to change if transhumanist goals were going to become a reality. The only way to push for that was to write and speak and agitate in the political arena.

And, even though he wished there were others who could do it, when it came to this kind of graft, no one else, it seemed, had his work ethic. It wasn't a life just anyone could live: spending all of your time advocating for a small movement with little hope of success (at least in the eyes of the vast majority of the public). Not just anyone had the nous to get national attention and headlines.

The US Transhumanist Party, which Zoltan founded, flounders in obscurity these days. Its current leadership has not been successful at grabbing attention in the way Zoltan is. Or even the way other transhumanist projects can be, like Transhuman Coin, a cryptocurrency created by Peter Xing, Charles Awuzie, Alyse Sue, and Avinash Singh, which made a big splash on social media and gathered a significant following (three times larger than the US Transhumanist Party on Twitter, for example, and larger than Humanity+, as of this writing).

After his 2016 campaign, Zoltan left the Transhumanist Party in what he considered to be good hands, those of a small-l libertarian friend, futurist writer, and Ayn Rand admirer, Gennady Stolyarov II, who became chairman. Zoltan gave up all control and essentially passed on his creation for someone else to remold—although he kept ownership of the trademark.

The Transhumanist Party without Zoltan is something quite different. In fact, it's sort of the opposite of what Zoltan had set up the party to be. He'd created a publicity vehicle centered on his personality, the rest of the transhumanist movement be damned. In contrast, Gennady is "not good at all" at getting media, but good at uniting "the different forces" within transhumanism, in Zoltan's words.

Gennady's strength has been in turning the party into an entity that receives support from across the transhumanist spectrum, leaving behind its divisive origins. Gennady admits that there are "many differences" between his approach to the party and Zoltan's, and that "during my tenure as Chairman, there has been more of a focus on internal organization."[7] He even authorized a second official name for the party, the Transhuman Party, to accommodate those who thought this was better than the Transhumanist Party. "Accordingly, the full name of our organization hereby becomes 'United States Transhumanist Party / Transhuman Party,'" Gennady wrote in an official missive:

> Members henceforth have the choice to refer to it as either the "Transhumanist Party" (the trademarked term) or the "Transhuman Party" (the non-trademarked term), using these terms either together or apart or interchangeably, as they please. We hope that these options will enable individuals to bypass arguments over the trademark and collaborate on substantive matters…[8]

Another example of Gennady's more academic and democratic approach is how he led the revision of Zoltan's Transhumanist Bill of Rights. In early 2017, the Transhumanist Party published an updated edition of the document originally delivered to the US Capitol by the Immortality Bus. Zoltan's Version 1.0

was written in 20 minutes by him alone, had seven punchy paragraphs, and was just 297 words. Version 2.0, spearheaded by Gennady, had multiple authors within the party, and went through a voting process with all party members, who could suggest and vote on amendments. The final document was almost 2000 words long, spending exactly 648 words defining the meaning of "sentient entities." It was published by *Wired*, which was a major media coup.

Gennady was great at creating processes that everyone in the small Transhumanist Party community could buy into. The question is: Where does it go from there? Unlike in the days of the Immortality Bus and Zoltan's wild stunts, today's Transhumanist Party has close to zero traction with the wider public, the media, or wealthy potential donors. Zoltan's fear, that without a charismatic figurehead like him the Transhumanist Party would become another fringe talking shop, has so far proven true.

Commenting in late 2019 on the person the party had chosen as its 2020 presidential nominee, Zoltan told me: "The poor guy seems like a decent human being, but he just hasn't been getting any press." That candidate, Ben Zion, was later expelled from the party, before the 2020 election. The party's statement disavowing him described how Zion had "repeatedly engaged in hate-filled tirades" and "expressed glee at the prospect of a violent death" for Donald Trump. The final straw was when Zion attempted to create a lab-grown hamburger from his own cells. The Transhumanist Party does not support cannibalism, since it is contrary to its "forward-thinking vision."[9]

Gennady & Co. are in it for reasons other than publicity. Those who devote time to the Transhumanist Party today do so because they're committed to ideas about the future, and they like participating in a forum dedicated to those ideas. Gennady believes that the major US political parties will one day "implode"—having proven themselves unable to address the challenges of the future—and the Transhumanist Party will be ready to step into the void, having stayed out of the fray and devoted itself to serious ideas. Therefore, the party can afford to grow "gradually."[10] This is a far cry from the urgent and disruptive political revolution that Zoltan sought to bring about.

Quite apart from media savviness, another factor is that while there are many passionate transhumanist believers out there, most of them need day jobs. They just can't devote the kind of time to the cause that Zoltan does, because he got rich first.

After Zoltan established the US Transhumanist Party in 2014, the excitement it generated led to the creation of small transhumanist parties all over the world. But, a few years later, many of those parties were totally defunct.

The US Transhuman National Committee, the organization set up in 2015 to counter Zoltan's influence—with hopes of becoming a rival to the US Transhumanist Party—also disbanded within a few years.

It's hard to keep the excitement going, with no money, and little hope of electoral success or press coverage, and without a charismatic attention-grabber at the helm. "Unless you're self-funded like I am, it's hard to do these things, because the real world takes over, and you have to eat," Zoltan admits. The fact that there was no one else to step up only added to his sense of mission.

He's become super-protective of his time so that more of it can be given to the cause. Even his marriage and family life are allowed to operate only within certain parameters. According to Zoltan, he hears a frequent refrain from Lisa: that they should go out more... spend more time with other families... have dinner with other couples... get to know new, interesting people.

"And I'm like, 'No. There's no dinners like that, unless it's something that's useful.' Because every single minute of my life is very important. I'm a machine that has to run. If the machine runs properly, we end up maybe in the White House or we end up doing something really great with our lives. And if the machine doesn't run properly, especially because of what we consider these traditional obligations to family—jeez, I'm not going to give up my life for that."

He already feels that family life burdens his brain. He wouldn't give up the family life that he's worked hard to create, but he doesn't want to take on more. He already has his mom living with him, two kids... "I hear all the time, philosophers' greatest regret is that they just had too many kids."

Perhaps as a reaction against his own super-strict upbringing— which, of course, he also rebelled against at the time—Zoltan takes a laissez-faire approach to parenting. He wants his kids to be able to do what they want, not what he wants.

I put it to him that he benefited in the end from his disciplined upbringing: The strict exercise regime he was forced into from a young age made him disciplined and able to pursue his goals, enduring hardship if need be—whether it was sailing, building houses 12 hours a day, or his single-minded focus on transhumanism. Zoltan doesn't disagree. But he firmly believes that his daughters' world will be very different from the world he grew up in, with robots doing all kinds of human jobs. Knowledge and skills will be available just by inserting a chip into your head. He once told an interviewer that he argued with Lisa over whether Eva and Isla should get piano lessons: "I don't see the point when they'll just be downloading them one day."[11] He went so far as to ponder, in a *Quartz* article, whether he could stop saving for his kids' college, as they might be able to have their education uploaded directly to their brains.[12]

Zoltan certainly doesn't lack involvement in his daughters' lives: He takes them to school, he's home with them a lot of the time, he posts a lot of pictures with his family on social. But I can't help but wonder if his lack of wanting to set goals for his kids comes from the fact that he's so focused on his own goals and devoting time to those.

He tells me: "The truth is my kids will benefit from my legacy (and wealth). They won't need to formally work much if they don't want...So I don't push them like my father pushed me...I just want them to be near me, happy, and healthy, and interested in the world and making it better...I hold myself to a much higher standard than I do my kids."[13]

***

In a different world, he would rather live a different life. If the aims of the transhumanist movement had already been accomplished, there's no question he would live differently. He might even spend more time with family and friends, like Lisa wants. But he's fighting a war.

"If transhumanism starts taking hold, like environmentalism took hold of the culture—then, all of a sudden, politicians will get elected into office, pressure will be put on government members...

"If we could get a cultural shift, then a huge watershed moment would happen where a flood of money would flow into these companies and...within five or ten years, you'd see a huge amount of change."

If governments don't start getting behind transhumanist science, then there could be a decade or more of difference between when the problem of death *actually* gets solved and when it *could* have been solved.

"Let's say, 2050 versus 2035. And in the meantime—like I've said, I've told you before—a billion people could die or maybe 800 million extra people could die."

He's not wrong on the math. Around 60 million people die from all causes globally every year. In 15 years, that's 900 million.

"It's really a tragedy when you don't take it as a day-to-day war that has already begun a long time ago. Every single day that we don't do the most we can, people die."

If Google (which has a life-extension-research arm called Calico) solves death in 2070, that will be too late for him and so many others, he says. The world can't wait that long.

And that's why, in a nutshell, a transhumanist social movement is so important to him—to bring forward the moment the end of death occurs. To pressure society so that it happens sooner. The science may get there without any help from activists. But the more public pressure and support there is, the more likely it happens quickly, and more people—including Zoltan—will be saved.

The good news was that the longevity industry, the area of research dedicated to lengthening lifespans, had grown exponentially in the years since Zoltan started banging his drum. Bank of America predicts longevity science will soon be a $600 billion industry. Zoltan claims credit for at least a little bit of that growth, thanks to all the publicity he's brought to the cause. But the war is far from won.

So, Zoltan is always looking for new fronts to open up in the publicity war he's waging. Transhumanist-themed wine, grown in his vineyards, is one idea. Starting his own think-tank is another. That's one reason why he signed up for a Master's in Ethics at Oxford. With better academic credentials, his public persona might have more credibility (his electoral prospects too). Maybe a Zoltan Istvan Institute of Transhumanism, publishing ethical arguments and policy position papers, is on the cards.

He's even looking beyond America for new places to run for election. Because his parents were born in Hungary, Zoltan and his children are entitled to EU citizenship—he could live and work in any one of 27 European countries right now.

He's always believed his popularity outside the US, and attitudes to transhumanism outside the US, are at least as important as the battle on the home front...

"I see the chessboard," he says. "I can see how I can do checkmate."

If life were different, he would live differently.

But... It isn't. Yet.

And, given the life that he has, Zoltan has no regrets.

"The only real question anyone ever has to ask themselves is, 'If you would do it all over again, would you start over?'... I wouldn't. Because I've had too many good things happen to me... All I want is time."

Whenever he takes time off, he says, it means that more people will die. He needs to give all the time to this fight that he possibly can. Every hour that he doesn't devote to the cause, more lives are lost.

He speculates: What if 100 hours that he spent surfing last year had instead been dedicated to pushing the movement forward? Maybe those 100 hours of work could have resulted in death ending a few days earlier, and three million lives being saved. That's the way he looks at it.

He just needs enough time to save everyone.

Notes

Unless otherwise noted, all quotes from Zoltan are from a series of interviews with the author during 2018 to 2020. Other sources are referenced below.

Prologue

1   Zoltan Istvan, "Should a Transhumanist Run for US President?" *Huffington Post*, October 8, 2014 (updated December 8, 2014), https://www.huffpost.com/entry/should-a-transhumanist-be_b_5949688.

Chapter 1

1   Author interview with Gabriella Gyurko Ashford, June 10, 2020.

2   Zoltan Istvan, "Brookings Resident Shares Sailboating Story," *Curry Coastal Pilot*, July 21, 2001, p. 1B.

3   *Playboy*, quoted on the cover of Zoltan Istvan, *The Futuresist Cure* (Rudi Ventures, 2019).

Chapter 2

1   Zoltan Istvan sailing journal, August 9, 1994. Reproduced in an email to Jennifer Hile, July 23, 1998, provided to author by Istvan.

2   Zoltan Istvan, "The Way," poem provided to author by Istvan, December 19, 2018.

3   Robin Lee Graham and Derek L.T. Gill, *Dove* (HarperPerennial e-book, 2012). There are no page numbers in this Kindle edition.

4   Zoltan Istvan journal quote, provided to author by Istvan, December 19, 2018.

5   YouTube video, "Zoltan Istvan Visits the Mareki Tribe in Vanuatu for the National Geographic Channel," posted by Istvan, August 27, 2017 (original video 2002), https://www.youtube.com/watch?v=02w7WZ0D3zA&t=4s.

6   Zoltan Istvan, Facebook post, January 10, 2020.

7   Zoltan Istvan, "Lightening Ship Gets Personal After Grounding," *Ocean Navigator*, August 20, 2003, http://www.oceannavigator.com/July-August-2003/Lightening-ship-gets-personal-after-grounding/.

8   Zoltan Istvan, "Brookings Resident Shares Sailboating Story," *Curry Coastal Pilot*, July 21, 2001, p. 3B.

9   Zoltan Istvan sailing journal, August 9, 1994. Reproduced in an email to Jennifer Hile, July 23, 1998, provided to author by Istvan.

10   Zoltan Istvan, "Environmentalists Are Wrong: Nature Isn't Sacred and We Should Replace It," *Lifeboat Foundation*, April 13, 2019, https://lifeboat.com/blog/2019/04/environmentalists-are-wrong-nature-isnt-sacred-and-we-should-replace-it (originally published at *Maven Roundtable*).

Chapter 3

1   Jennifer Burns, *Goddess of the Market: Ayn Rand and the American Right* (Oxford: Oxford University Press, 2009), p. 9.

2   Anne C. Heller, *Ayn Rand and the World She Made* (New York: Anchor Books, 2009), p. 245.

3   Zoltan Istvan journal quote, provided to author by Istvan, December 19, 2018.

4   "Cryonics: Life in Suspension," *X Factor*, no. 4 (London: Marshall Cavendish, 1997), pp. 85–9.

5   Ayn Rand, *The Fountainhead* (London: Penguin, 2007), pp. 710–12.

Chapter 4

1   Zoltan Istvan, dir., *Pawns of Paradise*. Privately screened for author, November 13, 2019.

2   Zoltan Istvan, "Vietnam Villagers Find Profit, Risk in Bomb Hunting," National Geographic Channel, January 7, 2004. Republished at http://www.ziventures.com/ NatGeoArticles.html.

3   YouTube video, "Vietnam Bomb Diggers: Dangerous Jobs." Posted by Zoltan Istvan, May 9, 2010 (original video, National Geographic Channel Presents segment, 2004), https://www.youtube.com/watch?v=vegzOgsVF1Y.

4   Quoted in Zoltan Istvan, dir., *Pawns of Paradise*. Privately screened for author, November 13, 2019.

5   Author interview with Lisa Memmel, March 29, 2020.

6   Author interview with Lisa Memmel, March 29, 2020.

7   Author interview with Lisa Memmel, March 29, 2020.

Chapter 5

1   Author interview with Lisa Memmel, March 29, 2020.

2   Zoltan Istvan, *The Transhumanist Wager* (Futurity Imagine Media, 2013), p. 4.

3   Zoltan Istvan, *The Transhumanist Wager* (Futurity Imagine Media, 2013), pp. 41–2.

4   Zoltan Istvan, *The Transhumanist Wager* (Futurity Imagine Media, 2013), pp. 53–4, 58.

5   Zoltan Istvan, *The Transhumanist Wager* (Futurity Imagine Media, 2013), p. 165.

6   Zoltan Istvan, *The Transhumanist Wager* (Futurity Imagine Media, 2013), p. 298.

7   Journal quotes by Zoltan Istvan, provided to author by email, December 19, 2018.

8   Zoltan Istvan, *The Transhumanist Wager* (Futurity Imagine Media, 2013), p. 69.

9   Zoltan Istvan email to author, December 19, 2018.

10 Zoltan Istvan, *The Transhumanist Wager* (Futurity Imagine Media, 2013), p. 175.

11 Zoltan Istvan, *The Transhumanist Wager* (Futurity Imagine Media, 2013), p. 188.

12 Zoltan Istvan, *The Transhumanist Wager* (Futurity Imagine Media, 2013), p. 188.

13 Zoltan Istvan, *The Transhumanist Wager* (Futurity Imagine Media, 2013), p. 201.

14 Zoltan Istvan, *The Transhumanist Wager* (Futurity Imagine Media, 2013), pp. 225–6.

15 Patrick Donovan, "Is Anyone Listening?" *The Age*, May 5, 2006, https://www.theage.com.au/entertainment/music/is-anyone-listening-20060505-ge28xl.html.

Chapter 6

1 Ayn Rand, quoted in Jennifer Burns, *Goddess of the Market: Ayn Rand and the American Right* (Oxford: Oxford University Press, 2009), p. 195.

2 Dylan Matthews, "Zoltan Istvan's Presidential Campaign Is Wacky as Hell. It's Also Necessary," *Vox*, September 18, 2015, https://www.vox.com/2015/9/18/9352117/zoltan-istvan-2016-campaign.

3 Zoltan Istvan email to author, July 28, 2018.

4 Quoted in Stephen Holden, "Crimes Against Property, as Protests," *New York Times*, June 21, 2011, https://www.nytimes.com/2011/06/22/movies/if-a-tree-falls-documentary-on-earth-liberation-front.html.

5 Zoltan Istvan, Facebook post, May 17, 2013.

6 Giulio Prisco, review of *The Transhumanist Wager* by Zoltan Istvan, *KurzweilAI*, May 15, 2013, https://www.kurzweilai.net/book-review-the-transhumanist-wager.

7 Ann Reynolds, "Cheat the End," review of *The Transhumanist Wager* by Zoltan Istvan, *Huffington Post*, June 6, 2013, https://www.huffpost.com/entry/cheat-the-end_b_3397188.

8   Zoltan Istvan, *The Anti-Deathist: Writings of a Radical Longevity Activist* (Rudi Ventures, 2020), p. 4.

9   Conversation (unrecorded) with author, July 29, 2018.

10 Zoltan Istvan, *The Anti-Deathist: Writings of a Radical Longevity Activist* (Rudi Ventures, 2020), pp. 50–1.

11 Zoltan Istvan, "I'm an Atheist, Therefore I'm a Transhumanist," *Huffington Post*, December 5, 2013, https://www.huffpost.com/entry/im-an-atheist-therefore-i_b_4388778.

12 Comment on Zoltan Istvan Facebook post, December 5, 2013.

13 Micah Mattix, "I'm an Atheist, Therefore I'm a Transhumanist," *American Conservative*, December 6, 2013, https://www.theamericanconservative.com/prufrock/im-an-atheist-therefore-im-a-transhumanist/.

14 Zoltan Istvan, Facebook post, December 7, 2013.

15 Andrew Sullivan, "Should Atheists Be Transhumanists?" *The Dish*, December 15, 2013, http://dish.andrewsullivan.com/2013/12/15/should-atheists-be-transhumanists/.

16 Zoltan Istvan, "AI Day Will Replace Christmas as the Most Important Holiday in Less Than 25 Years," *Huffington Post*, December 24, 2013, https://www.huffpost.com/entry/ai-day-will-replace-christmas_b_4496550.

17 "B.j. Murphy," Facebook post, December 19, 2013.

18 Chris T. Armstrong, email correspondence with author, July 24, 2022.

19 Rick Searle, "Betting Against *The Transhumanist Wager*," Institute for Ethics and Emerging Technologies, September 16, 2013, https://ieet.org/index.php/IEET2/more/searle20130916.

20 Aubrey de Grey, Facebook post, April 18, 2014.

21 Heather Wagenhals, "Zoltan Istvan Discusses How to 'Interpret Your Finances in the Future' on This Week's Show," *Unlock Your Wealth* Radio Show, February 14,

2014, https://directory.libsyn.com/episode/index/show/unlockyourwealth/id/2679025.

22 Zoltan Istvan, "Artificial Wombs Are Coming, but the Controversy Is Already Here," *Motherboard: Tech by Vice* (no date), https://www.vice.com/en/article/8qx8kk/artificial-wombs-are-coming-and-the-controversys-already-here.

23 Zoltan Istvan, "It's Time to Consider Restricting Human Breeding," *Wired.co.uk*, August 14, 2014, https://www.wired.co.uk/article/time-to-restrict-human-breeding.

24 Elise Hilton, "Restricting 'Human Breeding,' Wherein I Call Zoltan Istvan a Moral Idiot," *Acton Institute Powerblog*, August 15, 2014, https://blog.acton.org/archives/71625-lets-restrict-human-breeding-wherein-call-zoltan-istvan-idiot.html.

25 Ed Driscoll, "*Wired* Magazine Calls for Birth Panels," *PJ Media*, August 17, 2014, https://pjmedia.com/eddriscoll/2014/08/17/birth-panels-n258991.

26 Zoltan Istvan, "Some Atheists and Transhumanists Are Asking: Should It Be Illegal to Indoctrinate Kids with Religion?" *Huffington Post*, September 15, 2014, https://www.huffpost.com/entry/some-atheists-and-transhu_b_5814484.

Chapter 7

1 John Zerzan, email to Zoltan Istvan, quoted in Jamie Bartlett, *The Dark Net: Inside the Digital Underworld* (London: William Heinemann, 2014). It is not possible to identify page numbers in this Kindle edition.

2 Ross Andersen, "Q&A: A Proud Luddite on Steve Jobs' Legacy," *The Atlantic*, October 13, 2011, https://www.theatlantic.com/technology/archive/2011/10/q-a-a-proud-luddite-on-steve-jobs-legacy/246622/.

3 Zoltan Istvan, "A Stanford University Debate: Transhumanism vs. Anarcho-Primitivism," *Huffington*

Post, November 20, 2014, https://www.huffpost.com/entry/a-stanford-university-deb_b_6186982.

4   Jamie Bartlett, *The Dark Net: Inside the Digital Underworld* (London: William Heinemann, 2014). It is not possible to identify page numbers in this Kindle edition.

5   Yaron Brook, "The Chaos to Come—A Contested Election," *Yaron Brook Show* (podcast), September 22, 2020.

6   All of the comments quoted here appear under Zoltan's Facebook post announcing the Transhumanist Party, October 27, 2014.

7   Jamie Bartlett, "Meet the Transhumanist Party: 'Want to Live Forever? Vote for Me,'" *Daily Telegraph*, December 23, 2014, http://www.telegraph.co.uk/technology/11310031/Meet-the-Transhumanist-Party-Want-to-live-forever-Vote-for-me.html.

8   Zoltan Istvan, "The New American Dream? Let the Robots Take Our Jobs," *Motherboard: Tech by Vice*, February 13, 2015, https://www.vice.com/en/article/539j45/the-new-american-dream-let-the-robots-take-our-jobs.

9   Emily Ladau, quoted in Anthony Cuthbertson, "Exoskeletons v Wheelchairs: Disability Advocates Clash with Futurists over 'Offensive' Solution," *International Business Times*, April 14, 2015, https://www.ibtimes.co.uk/exoskeletons-vs-wheel-chairs-disability-advocates-clash-futurists-over-offensive-solution-1496178.

10 Comment on Zoltan Istvan Facebook post, April 4, 2015.

11 Leah Smith, "Fix Cracks Not Crips," Center for Disability Rights (no date), https://www.cdrnys.org/blog/disability-dialogue/the-disability-dialogue-fix-cracks-not-crips/.

12 Zoltan Istvan, "What If One Country Achieves the Singularity First?" *Motherboard: Tech by Vice*, April 21, 2015, https://www.vice.com/en/article/ypwkej/what-if-one-country-achieves-the-singularity-first.

13 Christopher J. Benek, quoted in Zoltan Istvan, "When Superintelligent AI Arrives, Will Religions Try to Convert

It?" *Gizmodo*, February 4, 2015, https://gizmodo.com/when-superintelligent-ai-arrives-will-religions-try-t-1682837922.

Chapter 8

1 Author interview with Lisa Memmel, March 29, 2020.

2 Zoltan Istvan, "Immortality Bus with Zoltan Istvan," *Indiegogo*, https://www.indiegogo.com/projects/immortality-bus-with-zoltan-istvan#/.

3 Rachel Edler, email correspondence with author, August 9, 2022.

Chapter 9

1 Zoltan Istvan, quoted in Daniel Sollinger, dir., *Immortality or Bust* (2019).

2 Jamie Bartlett, *Radicals: Outsiders Changing the World* (London: William Heinemann, 2017), Kindle edition, p. 42.

3 Hank Pellissier, "I Quit the USA Transhumanist Party. Why? Zoltan's Non-Inclusive Leadership," *Transhumanity.net*, September 22, 2015, https://transhumanity.net/i-quite-the-usa-transhumanist-pary-why-zoltans-non-inclusive-leadership/.

4 Lincoln Cannon, "Transhumanists Disavow Zoltan Istvan Candidacy for US Presidency," *ipetitions* (no date), https://www.ipetitions.com/petition/transhumanists-disavow-zoltan-istvan-candidacy.

5 Lincoln Cannon, "Transhumanists Should Disavow Zoltan Istvan Candidacy," October 2, 2015 (updated January 5, 2021), https://lincoln.metacannon.net/2015/10/transhumanists-should-disavow-zoltan.html.

6 Hank Pellissier, "15 Questions Zoltan Istvan Is Avoiding — Why? What Are the Answers?" *Transhumanity.net*, October 16, 2015, https://transhumanity.net/15-question-zoltan-istvan-is-avoiding-why-what-are-the-answers-opinion/.

7 Jamie Bartlett, *Radicals: Outsiders Changing the World* (London: William Heinemann, 2017), Kindle edition, pp. 43–4.

8   Zoltan Istvan, quoted in Jamie Bartlett, *Radicals: Outsiders Changing the World* (London: William Heinemann, 2017), Kindle edition, p. 44.

9   Giulio Prisco, "Zoltan and the Anti-Istvan Petition—Jethro or Zoe?" *Turing Church*, October 18, 2015, http://turingchurch.com/2015/10/18/zoltan-and-the-anti-istvan-petition-jethro-or-zoe/.

10 "The Transhumanist Declaration," *Humanity+* (no date), https://www.humanityplus.org/the-transhumanist-declaration.

11 Jeb Bush, quoted in Daniel White, "Here's a Ranking of Republican Reactions to Donald Trump from 'Sad!' to 'Yuuge,'" *Time*, May 18, 2016, https://time.com/4325178/donald-trump-republican-support/.

12 Lindsay Graham and Ted Cruz, quoted in Annie Karni, "No One Attacked Trump More in 2016 Than Republicans. It Didn't Work," *New York Times*, August 19, 2019, https://www.nytimes.com/2019/08/13/us/politics/trump-attacks-republicans-democrats.html.

13 Author interview with Lisa Memmel, March 29, 2020.

14 Max More, quoted in Daniel Sollinger, dir., *Immortality or Bust* (2019).

15 Zoltan Istvan, "Transhumanist Rights Are the Civil Rights of the 21st Century," *Newsweek*, April 30, 2016, https://www.newsweek.com/transhumanism-zoltan-istvan-civil-rights-21st-century-453884.

16 Kyle Cantrell, quoted in Zoltan Istvan, "I Visited One of the Largest Megachurches in the US as an Atheist Transhumanist Presidential Candidate—Here's What Happened," *Business Insider*, December 2, 2015, https://www.businessinsider.com/transhumanist-zoltan-istvan-visits-one-of-the-largest-megachurches-in-the-us-2015-11?r=US&IR=T.

17 "About the Church of Perpetual Life," *Church of Perpetual Life* (no date), https://www.churchofperpetuallife.org/about.

18 Bill Faloon, quoted in "Special Guests Jane Schastnaya, Polina Mamoshina, and Zoltan Istvan," Perpetual Life (YouTube channel), November 22, 2015, https://www.youtube.com/watch?v=rwpO9twA_Ps&t=1891s.

19 Martine Rothblatt, quoted in Jessica Roy, "The Rapture of the Nerds," *Time*, April 17, 2014, https://time.com/66536/terasem-trascendence-religion-technology/.

20 Jessica Roy, "The Rapture of the Nerds," *Time*, April 17, 2014, https://time.com/66536/terasem-trascendence-religion-technology/.

21 John McAfee, quoted in Anthony Cuthbertson, "US Elections 2016: John McAfee and Zoltan Istvan Debate Cybersecurity, Immortality and Sexbots," *International Business Times*, December 11, 2015, https://www.ibtimes.co.uk/us-elections-2016-john-mcafee-zoltan-istvan-debate-cybersecurity-immortality-sexbots-1532859.

22 Alex Jones, quoted in Daniel Sollinger, dir., *Immortality or Bust* (2019).

23 Zoltan Istvan, "Immortality Bus Delivers Newly Created Transhumanist Bill of Rights to the US Capitol," *Huffington Post*, December 21, 2015 (updated December 6, 2015), https://www.huffpost.com/entry/immortality-bus-delivers-_b_8849450.

Chapter 10

1 Jimmy Dore, "The Pro-Science Presidential Candidate," Rebel HQ (YouTube channel), July 27, 2016, https://www.youtube.com/watch?v=Pva1IM591p8.

2 Zoltan Istvan, "What It's Like to Counter-Protest Christians as an Atheist Demonstrator at Both Political Conventions," *Daily Dot*, July 31, 2016 (updated May 26, 2021), https://

www.dailydot.com/debug/atheist-transhumanist-rnc-dnc-demonstrations/.

3   Zoltan Istvan, "A Transhumanist Goes to the Conventions," *Motherboard: Tech by Vice*, July 28, 2016, https://www.vice.com/en/article/53d5xa/a-transhumanist-goes-to-the-conventions.

4   Aaron Sankin, "Majority of Americans Dislike Both Trump and Clinton as Interest in Third-Party Spikes Online," *Daily Dot*, May 24, 2016 (updated May 26, 2021), https://www.dailydot.com/debug/donald-trump-hillary-clinton-poll-record-unpopularity-third-party-searches/.

5   Anthony Cuthbertson, "Fake News Site Southend News Network Claims Full Responsibility for Donald Trump Victory," *Newsweek*, November 16, 2016, https://www.newsweek.com/fake-news-trump-clinton-pizzagate-paedophile-election-521797.

6   Mackenzie Weinger, "7 Pols Who Praised Ayn Rand," *Politico*, April 26, 2012, https://www.politico.com/story/2012/04/7-pols-who-praised-ayn-rand-075667.

7   Zoltan Istvan, quoted in Eric Mack, "The Surprising Way Transhumanist Zoltan Istvan Could Make It to the White House," *CNET*, May 4, 2016, https://www.cnet.com/news/the-surprising-way-transhumanist-candidate-zoltan-istvan-could-make-it-to-the-white-house/.

8   Eric Mack, "The Surprising Way Transhumanist Zoltan Istvan Could Make It to the White House," *CNET*, May 4, 2016, https://www.cnet.com/news/the-surprising-way-transhumanist-candidate-zoltan-istvan-could-make-it-to-the-white-house/.

9   Gary Johnson, email correspondence with author, October 20, 2022.

10 Steven Gyurko, quoted in Zoltan Istvan, "Futurist Zoltan Istvan Interviews His Aging Father," *Huffington Post*

(video), November 5, 2016 (updated December 6, 2017), https://www.huffpost.com/entry/talktome-futurist-zoltan-_b_9879872.

Chapter 11

1 Zoltan Istvan, Twitter direct message to author, November 11, 2019.

2 Zoltan Istvan, "Get a Grip on Reality, Libertarians—Start Winning Some Elections!" *Daily Caller*, February 5, 2018, https://dailycaller.com/2018/02/05/get-a-grip-on-reality-libertarians-start-winning-some-elections/.

3 Zoltan Istvan, Twitter direct message to author, February 16, 2020.

4 Comment on Zoltan Istvan Facebook post, November 19, 2019.

5 Zoltan Istvan, response to comment, Facebook, November 19, 2019.

6 Zoltan Istvan, Facebook post, February 15, 2020.

7 Zoltan Istvan, "Why I'm Running for California Governor as a Libertarian," *Newsweek*, February 12, 2017, https://www.newsweek.com/zoltan-istvan-california-governor-libertarian-555088.

8 Zoltan Istvan, Facebook post, January 8, 2020.

9 Zoltan Istvan, Facebook post, February 11, 2020.

10 Zoltan Istvan, Facebook post, February 12, 2020.

11 Zoltan Istvan, Facebook post, February 7, 2020.

12 "Zoltan Istvan's 20-Point Plan," quoted in a comment on Zoltan Istvan Facebook post, February 15, 2020.

13 Zoltan Istvan, Facebook post, February 15, 2020.

14 Author interview with Pratik Chougule, March 8, 2022.

15 Author interview with Pratik Chougule, March 8, 2022.

16 Zoltan Istvan, "The Transhumanism Movement Aims to Eliminate Existential Risk for the World," *Huffington*

Post, February 12, 2016 (updated December 6, 2017), https://www.huffpost.com/entry/the-transhumanism-movemen_b_9215418.

17   Zoltan Istvan, Facebook post, March 22, 2020.

18   Zoltan Istvan, "A Letter About Coronavirus, the Longevity Movement, & Why Quarantining Is Killing Us," *Medium*, March 24, 2020, https://zoltanistvan.medium.com/a-letter-about-coronavirus-the-longevity-movement-why-quarantining-is-killing-us-a4c1545eda9.

19   Zoltan Istvan, "Should Surfing Be Allowed During the Pandemic?" *New York Times*, May 1, 2020, https://www.nytimes.com/2020/05/01/opinion/surfing-coronavirus-quarantine.html.

20   Zoltan Istvan, Instagram post, May 4, 2020.

21   Zoltan Istvan, Twitter direct message to author, May 8, 2020.

22   Comment on Zoltan Istvan Facebook post, March 9, 2020.

23   Zoltan Istvan, Facebook post, March 8, 2020.

24   Comment on Zoltan Istvan Facebook post, March 8, 2020.

25   Zoltan Istvan, Facebook post, May 31, 2020.

26   Zoltan Istvan, Facebook post, June 1, 2020.

Chapter 12

1   Zoltan Istvan, *The Transhumanist Wager* (Futurity Imagine Media, 2013), pp. 283–4.

2   Lisa Nash, quoted in "17 Years Later, Nash Family Opens Up About Controversial Decision to Save Dying Daughter," *TheDenverChannel.com*, November 14, 2017, https://www.thedenverchannel.com/news/local-news/17-years-later-nash-family-opens-up-about-controversial-decision-to-save-dying-daughter.

3   Zoltan Istvan (@zoltan_istvan) tweet, November 14, 2021.

4   Zoltan Istvan (@zoltan_istvan) tweet, February 5, 2022.

5   Jacque Fresco, foreword to *The Futuresist Cure* by Zoltan Istvan (Rudi Ventures, 2019). It is not possible to identify page numbers in this Kindle edition.

6   Zoltan Istvan, email correspondence with author, March 3, 2022.

7   Gennady Stolyarov II, email correspondence with author, July 26, 2022.

8   Gennady Stolyarov II, "The U.S. Transhumanist Party / Transhuman Party: The Last, Best Hope for Transhumanist Politics," *United States Transhumanist Party*, January 26, 2019, https://transhumanist-party.org/2019/01/26/ustp-last-best-hope/.

9   United States Transhumanist Party, "U.S. Transhumanist Party Official Statement Regarding Johannon Ben Zion," June 24, 2020, https://transhumanist-party.org/2020/06/24/ustp-statement-ben-zion/.

10   Gennady Stolyarov II, "The U.S. Transhumanist Party / Transhuman Party: The Last, Best Hope for Transhumanist Politics," *United States Transhumanist Party*, January 26, 2019, https://transhumanist-party.org/2019/01/26/ustp-last-best-hope/.

11   Zoltan Istvan, quoted in Emily Gillingham, "Bodyhackers," *BAMM RAW Insights*, May 19, 2016, https://bammraw.com/new-blog/2016/5/19/bodyhackers.

12   Zoltan Istvan, "In 15 Years We'll Be Able to Upload Education to Our Brains. So Can I Stop Saving for My Kids' College?" *Quartz*, July 1, 2019, https://qz.com/1651749/brainwave-tech-will-make-college-debt-a-thing-of-the-past/.

13   Zoltan Istvan, email correspondence with author, March 3, 2022.

CHANGEMAKERS
BOOKS

Transform your life, transform our world. Changemakers
Books publishes books for people who seek to become
positive, powerful agents of change. These books
inform, inspire, and provide practical wisdom
and skills to empower us to write the next
chapter of humanity's future.

www.changemakers-books.com